Gamification and Advanced Technology to Enhance Motivation in Education

Gamification and Advanced Technology to Enhance Motivation in Education

Editors

**Faraón Llorens-Largo
Rafael Molina-Carmona**

MDPI • Basel • Beijing • Wuhan • Barcelona • Belgrade • Manchester • Tokyo • Cluj • Tianjin

Editors
Faraón Llorens-Largo
Smart Learning Research Group,
Department of Computer
Science and Artificial
Intelligence, University of
Alicante
Spain

Rafael Molina-Carmona
Smart Learning Research Group,
Department of Computer
Science and Artificial
Intelligence, University of
Alicante
Spain

Editorial Office
MDPI
St. Alban-Anlage 66
4052 Basel, Switzerland

This is a reprint of articles from the Special Issue published online in the open access journal *Informatics* (ISSN 2227-9709) (available at: https://www.mdpi.com/journal/informatics/special_issues/Gamification).

For citation purposes, cite each article independently as indicated on the article page online and as indicated below:

LastName, A.A.; LastName, B.B.; LastName, C.C. Article Title. *Journal Name* **Year**, *Article Number*, Page Range.

ISBN 978-3-03936-970-6 (Hbk)
ISBN 978-3-03936-971-3 (PDF)

© 2020 by the authors. Articles in this book are Open Access and distributed under the Creative Commons Attribution (CC BY) license, which allows users to download, copy and build upon published articles, as long as the author and publisher are properly credited, which ensures maximum dissemination and a wider impact of our publications.

The book as a whole is distributed by MDPI under the terms and conditions of the Creative Commons license CC BY-NC-ND.

Contents

About the Editors vii

Rafael Molina-Carmona and Faraón Llorens-Largo
Gamification and Advanced Technology to Enhance Motivation in Education
Reprinted from: *Informatics* 2020, 7, 20, doi:10.3390/informatics7020020 1

Daniel Corona Martínez and José Julio Real García
Using Malone's Theoretical Model on Gamification for Designing Educational Rubrics
Reprinted from: *Informatics* 2019, 6, 9, doi:10.3390/informatics6010009 7

Pedro C. Santana-Mancilla, Miguel A. Rodriguez-Ortiz, Miguel A. Garcia-Ruiz, Laura S. Gaytan-Lugo, Silvia B. Fajardo-Flores and Juan Contreras-Castillo
Teaching HCI Skills in Higher Education through Game Design: A Study of Students' Perceptions
Reprinted from: *Informatics* 2019, 6, 22, doi:10.3390/informatics6020022 21

Zoltan Buzady and Fernando Almeida
FLIGBY—A Serious Game Tool to Enhance Motivation and Competencies in Entrepreneurship
Reprinted from: *Informatics* 2019, 6, 27, doi:10.3390/informatics6030027 33

Oriol Borrás-Gené, Margarita Martínez-Núñez and Luis Martín-Fernández
Enhancing Fun through Gamification to Improve Engagement in MOOC
Reprinted from: *Informatics* 2019, 6, 28, doi:10.3390/informatics6030028 51

Marta Martín-del-Pozo, Ana García-Valcárcel Muñoz-Repiso and Azucena Hernández Martín
Video Games and Collaborative Learning in Education? A Scale for Measuring In-Service Teachers' Attitudes towards Collaborative Learning with Video Games
Reprinted from: *Informatics* 2019, 6, 30, doi:10.3390/informatics6030030 71

Alessandra Antonaci, Roland Klemke and Marcus Specht
The Effects of Gamification in Online Learning Environments: A Systematic Literature Review
Reprinted from: *Informatics* 2019, 6, 32, doi:10.3390/informatics6030032 85

Teresa Rojo, Myriam González-Limón and Asunción Rodríguez-Ramos
Company–University Collaboration in Applying Gamification to Learning about Insurance
Reprinted from: *Informatics* 2019, 6, 42, doi:10.3390/informatics6030042 107

Lizhou Cao, Chao Peng and Jeffrey T. Hansberger
Usability and Engagement Study for a Serious Virtual Reality Game of Lunar Exploration Missions
Reprinted from: *Informatics* 2019, 6, 44, doi:10.3390/informatics6040044 127

Francisco J. Gallego-Durán, Carlos J. Villagrá-Arnedo, Rosana Satorre-Cuerda, Patricia Compañ-Rosique, Rafael Molina-Carmona and Faraón Llorens-Largo
A Guide for Game-Design-Based Gamification
Reprinted from: *Informatics* 2019, 6, 49, doi:10.3390/informatics6040049 143

Teodora Iulia Constantinescu and Oswald Devisch
Serious Games, Mental Images, and Participatory Mapping: Reflections on a Set of Enabling Tools for Capacity Building
Reprinted from: *Informatics* 2020, 7, 7, doi:10.3390/informatics7010007 163

About the Editors

Faraón Llorens-Largo, Associate Professor: Dr Llorens-Largo is an associate professor of Computer Science and Artificial Intelligence of the University of Alicante (Spain). He is a graduate in Primary Education from the University of Alicante, and holds a degree in Computer Science from the Polytechnic University of Valencia and a Ph.D. in Computer Science from the University of Alicante. He has held various management positions, including Director of the Polytechnic School (2000–2005) and Vice-Rector of Technology and Educational Innovation (2005–2012), both at the University of Alicante, and Executive Secretary of the ICT Sector Commission of the Crue Universidades Españolas (2010–2012). He has received the Sapiens 2008 Prize for the Professional, awarded by the Official College of Computer Science Engineers of the Valencian Region, and the AENUI 2013 Prize for Teaching Quality and Innovation, awarded by the Association of University Teachers of Computer Science. He is the director of the "Cátedra Santander-UA de Transformación Digital" at the University of Alicante, devoted to exploring new trends in digital transformation; a member of the "Smart Learning" research group on Intelligent Technologies for Learning; and co-leader of GTI4U, an international team of researchers and IT professionals from different universities, whose objective is to research and promote the implementation of IT governance systems and digital transformation in universities. His work is in the fields of artificial intelligence, video game development and gamification, the application of digital technologies to education, IT governance and the digital transformation of universities.

Rafael Molina-Carmona, Associate Professor: Dr. Molina-Carmona received his B.Sc. and M.Sc. in Computer Science from the Polytechnic University of Valencia (Spain) in 1994, and his Ph.D. in Computer Science from the University of Alicante (Spain) in 2002. He is an Associate Professor at the University of Alicante, and he belongs to the department of Computer Science and Artificial Intelligence. His research interests and expertise lie in the areas of using information technologies to transform society and technological innovation. In particular, he works on the digital transformation of learning, innovation management and Artificial Intelligence applications in different fields: computer-aided design and manufacturing, computer graphics, learning, gamification, IT governance and information representation. He is the director of the "Smart Learning" research group on Intelligent Technologies for Learning: adaptive learning, learning analytics and predictive systems, gamification, video games and the digital transformation of educational institutions. He is also the director of the Laboratory of Technological Innovation in Education, belonging to the Scientific Innovation Unit of the University of Alicante, devoted to transferring innovations to industry; a member of "Cátedra Santander-UA de Transformación Digital" at the University of Alicante; and a member of GTI4U, an international team of researchers and IT professionals from different universities, aimed at researching and promoting the implementation of IT governance systems and digital transformation in universities. He has published more than 40 works, has made 60 contributions to conferences and has supervised seven theses in these fields.

Editorial

Gamification and Advanced Technology to Enhance Motivation in Education

Rafael Molina-Carmona [1,2,*] and Faraón Llorens-Largo [1,2]

1. Smart Learning Research Group, University of Alicante, 03690 Alicante, Spain; faraon.llorens@ua.es
2. Cátedra Santander-UA de Transformación Digital, University of Alicante, 03690 Alicante, Spain
* Correspondence: rmolina@ua.es

Received: 25 May 2020; Accepted: 19 June 2020; Published: 23 June 2020

Abstract: The aim of this Special Issue is to compile a set of research works that highlight the use of gamification and other advanced technologies as powerful tools for motivation during learning. We have been fortunate to obtain a representative sample of the current research activity in this field.

Keywords: gamification; serious games; motivation

1. Introduction

Motivation is the driving force behind many human activities, in particular learning. Motivated students are ready to make a significant mental effort and use deeper and more effective learning strategies. Some of the fundamental attributes of learning strategies that enhance motivation are:

- Experimentation or learning by doing.
- Interactivity and immediate feedback.
- Allow and naturalize the error.
- Give control to the learner.

Numerous studies indicate that playing promotes learning, since when fun pervades the learning process, motivation increases, and tension is reduced. There is no doubt that the games comply with the four learning strategies that we have defined: playing is action and, therefore, experimentation and active learning; without interaction and continuous feedback, there is no game either; trial and error is the fundamental element thanks to which people learn to play and improve; and it is the player who maintains control of the game and who decides what actions to take at each step.

From our point of view, games can be very powerful tools in the improvement of learning processes from three different and complementary perspectives: as tools for teaching content or skills, as an object of the learning project itself, and as a philosophy to be taken into account when designing the training process. Each contributions presented in this Special Issue "Gamification and Advanced Technology to Enhance Motivation in Education" falls into one of these categories, that is to say, they all deal with the use of games or related technologies, and they all study how playing enhances motivation in education. In the following sections, we present the papers by establishing a thread that provides integrity to the Special Issue.

2. Games as Teaching Tools

The first group of papers consists of five contributions, in which the use of games to train students in different contents or skills is proposed. The first three papers present three serious games as tools for the teaching of subjects related to economic activity and the history of lunar explorations, in all cases with the motivation of the students as the main focus. In the fourth article, a serious game is complemented by other technological tools, in this case to develop the collaborative capacities of a population for the urban transformation of its environment. The last article in this section has a somewhat different, but related objective, as it attempts to analyze the attitude of teachers towards the use of games in the classroom.

The work of Teresa Rojo, Myriam González-Limón, and Asunción Rodríguez-Ramos [1] studies the use of a serious game, BugaMAP, for university teaching about insurance and evaluates its potential both objectively and subjectively. To do so, the authors propose a student opinion questionnaire evaluating the BugaMAP game (subjective evaluation); and the analysis of the patterns of the gamification of learning contained in the game (objective evaluation). Several variables are considered about aspects such as narrative, decision-making, short-term and long-term objectives, gifts for efforts, quick and clear responses, uncertainty, assistance, learning from decision impacts, and interaction with other players. Very valuable conclusions are obtained: a high satisfaction of the students with the knowledge acquired using fun and social interaction; the important role of the university professors and the company monitors; and the benefits of the company-university collaboration.

Zoltan Buzady and Fernando Almeida [2], for their part, present a serious game as a tool to acquire skills and abilities. FLIGBY is a serious game, which allows students to develop entrepreneurship skills in an immersive way based on real challenges that can be found in business environments. The authors propose a simple, but very effective quantitative approach to assess the success of FLIGBY adoption. They conclude that by using FLIGBY, students are able to train their skills in a wide range of domains like gathering information, motivating employees, training their emotional intelligence, and establishing social dynamics in a corporate environment. Two benefits are highlighted in the paper: (1) this informal teaching method based on a serious game allows students to increase technical skills in the field of management and entrepreneurship and also allows them to develop essential soft-skills; and (2) the authors have detected the arousal in other higher education institutions of the desire to include serious games as a complementary activity to formal teaching methods.

Creating a serious game about a mission to the moon and evaluating the use of Virtual Reality (VR) with respect to usability and engagement are the focus of Lizhou Cao, Chao Peng, and Jeffrey T. Hansberger's contribution [3]. In this work, they design and implement a serious VR game that immerses players into activities of lunar exploration missions in a virtual environment. They study the usability and engagement of the game through user experience in both VR and non-VR versions of the game, through the use of the Game Engagement Questionnaire (GEQ) and an interview questionnaire to measure levels of engagement. The experimental results show that the VR version of the lunar roving game took longer for participants to finish, but enhanced the game engagement and their motivation to learn the events of lunar exploration.

The work of Teodora Iulia Constantinescu and Oswald Devisch [4] also analyzes the use of serious games (and other technological tools), but in this case, with a different nuance. Specifically, the authors study how these tools enable collaboration between groups, in other words, how they contribute to training people to be able to collaborate, all within the context of actions to transform an urban area. To do so, the authors propose a conceptual framework for building capacities, in which the process and outputs collide with the ideas of choice, ability, and opportunity. The case study looks at one of the main commercial streets of the city of Ghent, Belgium (Vennestraat), and reflects on a set of enabling artifacts used to engage proprietors in the capacity-building process. This capacity-building process, characterized by the idea of space and capabilities, advances a critical viewpoint on issues related to participatory processes and gives practitioners a set of enabling tools to start a conversation about complex urban transformations.

Finally, in this section, Marta Martín-del-Pozo, Ana García-Valcárcel Muñoz-Repiso, and Azucena Hernández Martín [5] propose the creation of an attitude scale that primary school teachers present towards the use of video games for collaborative learning. In this case, a specific game is not used as a case study, but rather the opinion of the teachers is sought. The authors argue that for games to be successfully introduced into children's education, it is essential that teachers themselves are motivated to use them. Therefore, they propose to measure the attitude towards these new methodologies among the teachers, by creating an attitude scale towards collaborative learning with video games. They follow different methodological steps to make the scale construction possible, such as the analysis of items and the verification of their reliability, resulting in a rigorous attitude scale of 33 items, with high reliability, so that the measurement instrument can be considered as useful and valid.

3. Games as Learning Objects

Games can also be the object of the learning process. Although this aspect is better suited to some disciplines than others, there are more and more areas in which the study of games is not only interesting, but also convenient. This is undoubtedly true in the case of computer engineering and similar fields that can use video games as projects to be developed by students. A different case, but also very interesting, is the analysis of games to study aspects that are an intrinsic part of video games, such as their interaction, usability, visual design, or the artificial intelligence of their characters. This is the case of the work of Pedro C. Santana-Mancilla, Miguel A. Rodriguez-Ortiz, Miguel A. Garcia-Ruiz, Laura S. Gaytan-Lugo, Silvia B. Fajardo-Flores, and Juan Contreras-Castillo [6], which studies aspects of Human-Computer Interaction (HCI) through the analysis of video games. From our point of view, this paper makes two key contributions: a proposal for using the design and evaluation of computer games as a learning tool to teach HCI to undergraduate students; and the empirical validation of this proposal to explore students' attitudes with the aim to understand if the students believe that using video games allows them to learn higher education skills. The experimental results of the validation process indicate that using video games as teaching method provides the students with the HCI skills (psychology of everyday things, involving users, task-centered system design, models of human behavior, creativity and metaphors, and graphical screen design), and more importantly, they have a positive perception of the efficacy of the use of video game design in a higher education course.

4. Games as a Design Philosophy: Gamification

Finally, games can serve as inspiration for the development of teaching methodologies that, without being exactly games, incorporate the philosophy of games, both in their design and in their learning objectives. This is the case of gamification, which has been gaining momentum in recent years. Gamification consists of applying the principles of video game design, the use of mechanics and the elements of a game in any process, beyond the context of video games. The aim is to take advantage of both the psychological predisposition of people to participate in games and the quality of the game to motivate and improve the behavior of the participants. Precisely for this reason, due to its motivating nature, it is the subject of analysis on four of the contributions in this Special Issue, each with a different point of view. These include a review of the literature on the effects of gamification on online learning, a case study of gamification in a Massive Online Open Course (MOOC), a guide to the gamification of learning activities, and a methodology for designing assessment rubrics inspired by the principles of gamification.

A good way to understand a topic is to review the corresponding literature. That is what Alessandra Antonaci, Roland Klemke, and Marcus Specht [7] do in their systematic review of the literature on the effects of gamification in online learning environments. In this study, they identify 24 gamification elements that, combined, produce empirical effects on users' behavior in online learning in six areas: performance, motivation, engagement, attitude towards gamification, collaboration, and social awareness. This contribution is significant, and it is reinforced by the fact that the other papers presented in this Special Issue are in one of these six areas. The findings of this literature review point out that gamification and its application in online learning and in particular in MOOCs are still a young field, lacking empirical experiments and evidence, with a tendency of using gamification mainly as external rewards.

The second article in this section, by Oriol Borrás-Gené, Margarita Martínez-Núñez, and Luis Martín-Fernández [8], fills part of the gap detected in the previous literature review, presenting a particular case of the introduction of gamification elements in an MOOC and its associated Virtual Learning Community (VLC), together with a study on the fun and engagement that are achieved through gamification. The aim of this research is to find out whether, through the application in one MOOC, with a connectivism approach, of various gamification techniques, which increase motivation and fun, it is possible to achieve a greater engagement in terms of participation and generate a habit in the use of the VLC. To do so, a satisfaction survey, based on the validated SEEQsurvey, is conducted. The results show an increment of active participation and engagement within the MOOC community in the form of content creation and, especially, greater interaction and the generation of a habit so that the activity continues once the edition of the MOOC is finished.

The contribution of Francisco J. Gallego-Durán, Carlos J. Villagrá-Arnedo, Rosana Satorre-Cuerda, Patricia Compañ-Rosique, Rafael Molina-Carmona, and Faraón Llorens-Largo [9] emphasizes the relationship between video games and gamification. The authors consider that if there are video games that entertain and engage players in a very remarkable way, it is interesting to study what the design principles of these games are to try to bring them into a gamification process. Although there is no consensus on these design principles and much of the success of some video games is due to the experience and know-how of game designers, it is possible to establish guidelines that can help design more effective, challenging, and engaging gamification experiences. The guidelines are presented in the form of a rubric for educators and researchers to start working in gamification without previous experience in game design. This rubric decomposes the continuous space of game design into a set of ten discrete characteristics, based on the previous design experience of the authors, compared and contrasted with the literature, and empirically tested with some example games and gamified activities. As a consequence, a better understanding of the strengths and weaknesses of gamification and some tips to help in the design or improvement of activities are obtained.

The last article in the Special Issue is by Daniel Corona Martínez and José Julio Real García [10]. In their work, gamification is used as an inspiration to develop assessment rubrics. They make a methodological proposal to design educational rubrics to assess students' experiences of the active methodologies in secondary education. Their goal is to design better educational rubrics based on Malone's game theory and theoretical models of game design. As a main contribution, they propose a translation from game theory ideas into didactical concepts that can be transcribed and used in a didactical approach. As a consequence, they propose five main concepts that appear constantly and repeatedly during Malone's argumentation: challenge, curiosity, fantasy, design, and environment. These concepts are used to define the rubric, but instead of a simple translation, the authors apply a didactic and educational filter prior to proceeding, resulting in the five items that are part of the rubric: achievement, originality, motivation, design and quality, and relationships and time management. The resulting evaluation rubric includes a holistic approach to all different aspects related to the evaluation for active methodologies in a secondary education environment.

5. Conclusions

Playing is a pleasant, motivating human activity from which much can be learned. Teachers can take advantage of the characteristics of games to impregnate their teaching methodologies and motivate their students. The interactive features of the games, the action, and the speed strengthens the neurons and links that are involved in the correct prediction by means of endorphins and dopamine, giving the player the sensation commonly known as fun. This is how learning and intrinsic motivation occur.

Motivation is one of the key aspects of good instructional design and becomes more important as the learner takes ownership of the process. In traditional teaching, the teacher in the classroom can react to the attitudes of the students. However, in online teaching, the interaction between teacher and students and students among themselves becomes less intense, and the interaction of students with learning resources becomes more important. In this sense, designing the learning experience with the aim of motivating the students will serve both face-to-face and remote teaching.

With the conviction that games, gamification, and other related technologies have this motivating potential, the call for participation in this Special Issue was made. The selected articles present innovative ideas, models, approaches, technologies, reviews, and case studies that contribute to creating a publication of high interest. The Guest Editors of this Special Issue would like to thank the authors for their great work, which enriches research on gamification and advanced technology to enhance motivation in education.

Author Contributions: Conceptualization, R.M.-C. and F.L.-L.; writing, original draft preparation, R.M.-C. and F.L.-L.; writing, review and editing, R.M.-C. and F.L.-L. All authors read and agreed to the published version of the manuscript.

Acknowledgments: The Special Issue Editors would like to acknowledge the reviewers for their essential contribution to the quality of the papers.

Conflicts of Interest: The authors declare no conflict of interest.

References

1. Rojo, T.; González-Limón, M.; Rodríguez-Ramos, A. Company–University Collaboration in Applying Gamification to Learning about Insurance. *Informatics* **2019**, *6*, 42. [CrossRef]
2. Buzady, Z.; Almeida, F. FLIGBY—A Serious Game Tool to Enhance Motivation and Competencies in Entrepreneurship. *Informatics* **2019**, *6*, 27. [CrossRef]
3. Cao, L.; Peng, C.; Hansberger, J.T. Usability and Engagement Study for a Serious Virtual Reality Game of Lunar Exploration Missions. *Informatics* **2019**, *6*, 44. [CrossRef]
4. Constantinescu, T.I.; Devisch, O. Serious Games, Mental Images, and Participatory Mapping: Reflections on a Set of Enabling Tools for Capacity Building. *Informatics* **2020**, *7*, 7. [CrossRef]
5. Martín-del Pozo, M.; García-Valcárcel Muñoz-Repiso, A.; Hernández Martín, A. Video Games and Collaborative Learning in Education? A Scale for Measuring In-Service Teachers' Attitudes towards Collaborative Learning with Video Games. *Informatics* **2019**, *6*, 30. [CrossRef]
6. Santana-Mancilla, P.C.; Rodriguez-Ortiz, M.A.; Garcia-Ruiz, M.A.; Gaytan-Lugo, L.S.; Fajardo-Flores, S.B.; Contreras-Castillo, J. Teaching HCI Skills in Higher Education through Game Design: A Study of Students' Perceptions. *Informatics* **2019**, *6*, 22. [CrossRef]
7. Antonaci, A.; Klemke, R.; Specht, M. The Effects of Gamification in Online Learning Environments: A Systematic Literature Review. *Informatics* **2019**, *6*, 32. [CrossRef]
8. Borrás-Gené, O.; Martínez-Núñez, M.; Martín-Fernández, L. Enhancing Fun Through Gamification to Improve Engagement in MOOC. *Informatics* **2019**, *6*, 28. [CrossRef]
9. Gallego-Durán, F.J.; Villagrá-Arnedo, C.J.; Satorre-Cuerda, R.; Compañ-Rosique, P.; Molina-Carmona, R.; Llorens-Largo, F. A Guide for Game-Design-Based Gamification. *Informatics* **2019**, *6*, 49. [CrossRef]
10. Corona Martínez, D.; Real García, J. Using Malone's Theoretical Model on Gamification for Designing Educational Rubrics. *Informatics* **2019**, *6*, 9. [CrossRef]

© 2020 by the authors. Licensee MDPI, Basel, Switzerland. This article is an open access article distributed under the terms and conditions of the Creative Commons Attribution (CC BY) license (http://creativecommons.org/licenses/by/4.0/).

Article

Using Malone's Theoretical Model on Gamification for Designing Educational Rubrics

Daniel Corona Martínez * and José Julio Real García

Department of Didactics and Educational Theory, Research Group: Educational Research for the Transformation of Education (GICE-UAM), Universidad Autónoma de Madrid, Madrid 28049, Spain; julio.real@uam.es
* Correspondence: danielcormar@gmail.com

Received: 28 January 2019; Accepted: 27 February 2019; Published: 4 March 2019

Abstract: How could a structured proposal for an evaluation rubric benefit from assessing and including the organizational variables used when one of the first definitions of gamification related to game theory was established by Thomas W. Malone in 1980? By studying the importance and current validity of Malone's corollaries on his article *What makes things fun to Learn?* this work covers all different characteristics of the concepts once used to define the term "gamification." Based on the results of this analysis, we will propose different evaluation concepts that will be assessed and included in a qualitative proposal for an evaluation rubric, with the ultimate goal of including a holistic approach to all different aspects related to evaluation for active methodologies in a secondary education environment.

Keywords: gamification; active methodologies; secondary education; evaluation rubric; evaluation criteria; Thomas W. Malone; game; design; Sebastian Deterding; Nick Pelling

---

1. Introduction

Gamification (for many students and even for their teachers) seems to be a very enjoyable learning method, since it consists of playing and having fun while learning, so it is a facet of learning encouraged by all levels of the educational system. The gamification model has been used for a long time by various companies, according to the website Wonnova [1], which applies this model to the movie Jumanji.

The corporate idea of gamification has been used to motivate workers and customers to perform certain actions, for example, to define certain objectives with a motivational goal, to design a plan, to create game mechanics within the business environment, to develop a sense of belonging, or to design a company's motto.

From an educational point of view, gamification is not a new strategy, as game concepts have been used to motivate, stimulate, and impart content ever since the schooling system's early times. The use of these game elements will not only be performed during the early childhood educational stages, as these playful experiences can be enjoyed regardless of age and whether or not it is clear that present educational elements are embedded on these strategies.

Another aspect that should not be forgotten within this educational aspect of the concept is its possible influence over the evaluative stage of learning. When performing different methodological strategies, it is crucial to evaluate them in different ways. There are various recent examples for these "alternative" evaluations [2,3]. As an example of other papers covering this topic, Munuera and Ruiz express that the use of these concepts related to game theory could "allow [one] to generate a motivating environment to knowledge, as you can instantly know the evaluation of the contents and [the] student´s level of assimilation of them" [4].

In addition, it is interesting to state that there are some controversies among authors as to whether gamification will be framed within the context of behaviorism's paradigm. According to Borrás [5], the most dangerous theoretical aspects that behaviorist gamification could present are as follows:

1. *Manipulation*: In some gamification approaches, some ideas have clearly been put in place to influence students to choose specific paths, distorting students' right to free will (i.e., the idea of providing better or worse badges depending on the chosen strategy within a gamified activity).
2. *Hedonic Treadmill Reinforcement*: This idea poses a serious risk to more idealistic approaches to the gamification concept. There is a risk that, if individuals only act when there are rewards, they will reach a point where they might not continue playing if there are none. Based on this approach, it appears to be evident that it will be necessary to avoid using these gamification activities where the only goal is obtaining rewards, essentially because, by doing so, students might lose their motivation, the pleasure of obtaining a greater reward, or their will to fulfill a greater objective after obtaining these immediate gratifications.
3. *Overemphasis on Status*: Our state or position with respect to others is a very motivating element, as we, as humans, carry out actions to improve our own status. However, if the system only focuses on these elements, it can lead to demotivation (i.e., knowing that we will never be able to reach the first position in a gamified activity). Even more pertinently, many people do not feel this need of being recognized. This is a common error in gamification, as teachers will not focus on status alone.

The real methodological danger is over-focusing on prize searching, as this could deeply distort students' whole approach to learning process. According to what has been expressed (and taking into consideration that one of the main theoretical objectives in this paper is to avoid all commodification and reification related to gamification), a series of research questions have to be described and discussed throughout this paper:

- Could learning be improved by using gamification strategies?
- If so, how will this be reflected in students' performance?
- What is the cost of this learning? Can teachers take advantage of a new proposed evaluation method?

The main objectives of this work are clearly linked with the three questions expressed above. Based on our teaching experience, there will be a methodological proposal to design educational rubrics to assess students' experiences of the active methodologies in secondary education. By proposing this model, we intend to justify the advantages of gamification when it comes to motivating students and thus achieving an applicable and meaningful learning process for all parts involved. This paper will start with some common background related to the term "gamification," to then continue with the proposed method. After presenting it, there will be a discussion where we will discuss the implications of a holistic evaluation for the students involved in the process, followed by the conclusions.

2. Background

Over the years, there have been many different attempts to define the term "gamification." Although this paper will cover some of the most consistent attempts to define this idea, it has been hard to agree on a certain definition that will cover all of its complexity. Besides this, the term has been gaining *momentum* over the years, as more and more teachers and educational academics have been focusing on understanding and explaining how this specific term could be used to help to bridge the gap between a more conservative approach to education and the rise of active methodologies and its epistemological approach to teaching.

Based on the information stated in Figures 1 and 2 from Google Trends [6], the rise on the interest in the educational field began in September 2010 with the first significant result over 1 search per month. Since then, search results started to rise. This search result remains steady in an average of

75 searches per month, always descending during school vacation periods (Summer and Christmas). When a search term is compared with others from the same educational innovation field, such as project-based learning, the results are quite similar, even on their distribution throughout the year.

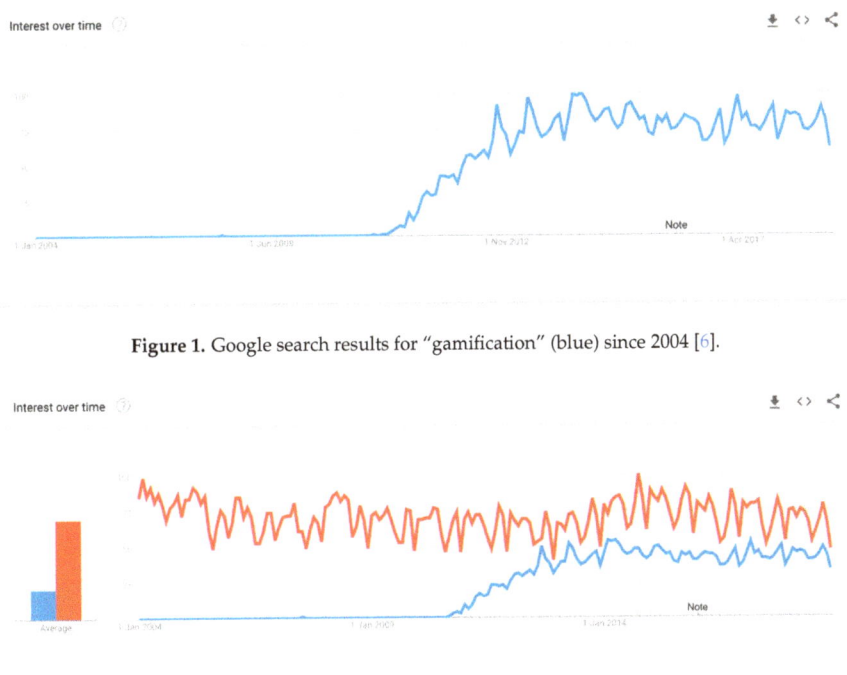

Figure 1. Google search results for "gamification" (blue) since 2004 [6].

Figure 2. Google search results for "gamification" (blue) vs. "project-based learning" (red) since 2004 [6].

Even though project-based learning has appeared in internet searches since the very beginning of Google Trends' data collection, it seems as though gamification has not aroused interest until late into the 2010s (in fact, the first time that the term reached 25 searches per month was in December 2011). It seems as though the interest in this discipline has a clear connection with some of the most relevant moments in the history of gamification's definitions and most of its trending moments. To ensure a proper understanding in the development of the term and its didactical sphere of influence, the following classification will provide a more in-depth description of gamification's most relevant and influential authors:

1) *September 1980*: Thomas W. Malone at Palo Alto Research Center wrote a slightly revised version of his Ph.D. dissertation submitted to the Department of Psychology at Stanford University. In his document *What makes Things Fun to Learn? A Study of Intrinsically Motivating Computer Games* [7], he defines various concepts related to game theory that later on will be used and recycled by other authors to define the term "gamification." Even when Malone did not coin the term, he made all the theoretical work to ensure that a new innovative framework could be generated where motivation would be at the central spot for computer game's users and game theorists. This work will be discussed in-depth throughout this article.

2) *Late 2002*: Nick Pelling, an English Computer Engineer and Game Developer coins the term, initially with the goal of developing a new way of dealing with transactions and activities on commercial electronic devices. In Pelling's own words, the goal was to "apply game-like

accelerated user interface design to make electronic transactions both enjoyable and fast." [8]. Even though the coining of its definition did not boost users' interest, this first definition was crucial for the later development of the discipline.

3) *March 2011*: Two months before Sebastian Deterding, defined by some authors as "one of the most influential thought leaders in the area of gamification" [9], Zimmerman & Cunningham defined gamification as "the process of game-thinking and game mechanics to engage users and solve problems." [10]. It was a significant definition as it shows, for one of the first times in history, the relation between game-thinking and problem-solving.

4) *May 2011*: One of the most relevant papers related to the actual definition of gamification was Deterding et al.'s *From Game Design Elements to Gamefulness: Defining "Gamification,"* which in May 2011 provided one of the most shared definitions of gamification, stating the term as "the use of game design elements in non-game contexts" [11], receiving broad support with 102 citations on June 16, 2013 [12]. One of the reasons why Deterding's paper became so influential is thanks to two specific factors: first because of the simplicity of his definition and second because of its good assessment of where to place the term in the *game* vs. *play* conceptual framework, where gamification is shown in a partially-related-to-gaming sphere of influence; furthermore, the author uses a very graphic explanation to describe its definition [11]. This made Deterding et al.'s definition the most widespread definition still in use.

5) *August 2011*: Finally, on this race to define (and own) the growing influence of this term, research and advisory American founded company Gartner defined gamification as one of the emerging technologies for their Hype Cycle in 2011 [13]. This tool is meant to provide help to strategists and planners with an assessment of the maturity, business benefit, and future direction of some of the top technologies in the world [14]. As can be seen in the next image, gamification is highlighted as one of the most interesting emerging technologies. Based on Dicheva et al., in 2013, the expectation for reaching the productivity plateau was five to ten years [15]. This position, however, reflected mainly its use in business contexts. Penetration of the gamification trend in educational settings seemed still to be becoming increasingly substantial, as indicated by the amount and annual distribution of reviewed works.

As exposed, and based on the evidence from past papers, some of the most relevant moments were in 2011, when three of the most relevant and influencing authors came together, boosting the expectancy related to the term and its academic and business use.

The reason why, in this paper, focus will be put on Malone's theories despite others with higher repercussion is simple. First, many other authors have tried to define gamification in different fields such as online software, idea competitions, citizen science, marketing, and many more [16–20], but most of these have focused on defining the concept from an economically driven point of view, as their proposals were deeply related to game design and user experiences in video games and computer science [15].

However, the term has been gaining much more relevance in fields such as consumer marketing and computer user interfaces. This has led to a *businessification* of the term "gamification," leaving the term empty of its educational goal in many ways. Exactly when the term started becoming increasingly relevant in the educational field, gamification as a concept was increasingly used in fields with a growing economy and plausible investment implications. This has led to a situation where the concept often "feels" out of tune with its original game-theory-related meaning. It has become a floating signifier. Even in actual education, the term seems to be defined as a floating signifier, as it is sometimes being used to push economical goals under the educational innovation idea framework.

This is not the case of Thomas W. Malone's Ph.D. dissertation objectives. In his thesis, Malone's goal is to find a game theory (and, of course, a model for that theory) that will allow games to be playful, desirable, motivational, fun to play, and enjoyable. The most relevant part of Malone's thesis for this paper is how these ideas could fit into a more educational approach to the term. In fact, after reviewing the most significant contributions throughout history, the various reasons why this article will be focusing on Malone's theoretical model are reinforced, especially from a modeling point of view.

By proposing a methodology whose main goal is to design better educational rubrics (based on Malone's game theory and theoretical models of game design) to assess students' experiences of the active methodologies in secondary education, the final objective is to realize the gamification concept with a more didactical (and enjoyable) meaning and to help to apply these ideas in modern innovation in education.

Apart from this main goal, other secondary goals on this paper are as follows:

a) to ensure that a clear review of the term "gamification" and its origins is taken into consideration when proposing a new evaluation rubric;
b) to use Malone's ideas and their relation to cover, from a holistic point of view, all aspects related to the evaluation of classes of active methodologies;
c) to propose a translation from game theory ideas into didactical concepts that can be transcribed and used in a didactical approach;
d) to discuss and propose the expected results of the theoretical experiment so as to increase their use for educational purposes;
e) to make *conceptual valuable enjoyment* the central paradigm of the gamification ideal.

3. Method

In order to make sure that Malone's ideas in *What makes Things Fun to Learn? A Study of Intrinsically Motivating Computer Games* [7] are used in a way that will help to build a new didactic paradigm related with evaluation rubrics, this article will present a conceptual map based on all of the ideas (and connections between them) that the author used in the revised version of this Ph.D. dissertation. After reviewing all main concepts, the most relevant concepts from Malone's theory will be introduced to design an educational rubric that can be used to assess and evaluate any kind of active methodology experience for students in different educational environments.

To clarify all concepts related to Malone's theory, a conceptual map will be presented in order to increase understanding of Malone's brilliant Ph.D. dissertation.

Based on Figure 3's conceptual map, there are clearly five main concepts that appear constantly and repeatedly during Malone's argumentation: *challenge, curiosity, fantasy, design,* and *environment*.

These will be the five concepts that will be used to define the rubric. Of course, this will not be done by simply translating these concepts and/or copying them into it, as a didactic and educational filter must be applied prior to proceeding. In order to make this possible and to make it an appealing tool for any teacher interested in using it, it is crucial to define which definition of an educational rubric will be used in this paper.

A rubric could be defined as "a scoring guide used to evaluate the quality of students' constructed responses" [21]. It is clearly an attempt to set expectations and to define an evaluation structure to advise the student in advance or during an assessment activity. In order for the teacher to do so, the rubric must contain at least two dimensions that will be used to judge students' performance, each of them with definitions and/or examples that will clarify the scale of values that will be used to evaluate each dimension. Additionally, standards of excellence for specified and holistic performance levels will be included. The most typical example of the rubric has the shape of a grid.

Once all definitions have been settled, the evaluation rubric based on Malone's conceptual map can be defined. Taking into consideration all variables, there will be two dimensions: one based on all the different holistic criteria on which students will be evaluated and another where performance levels will be included.

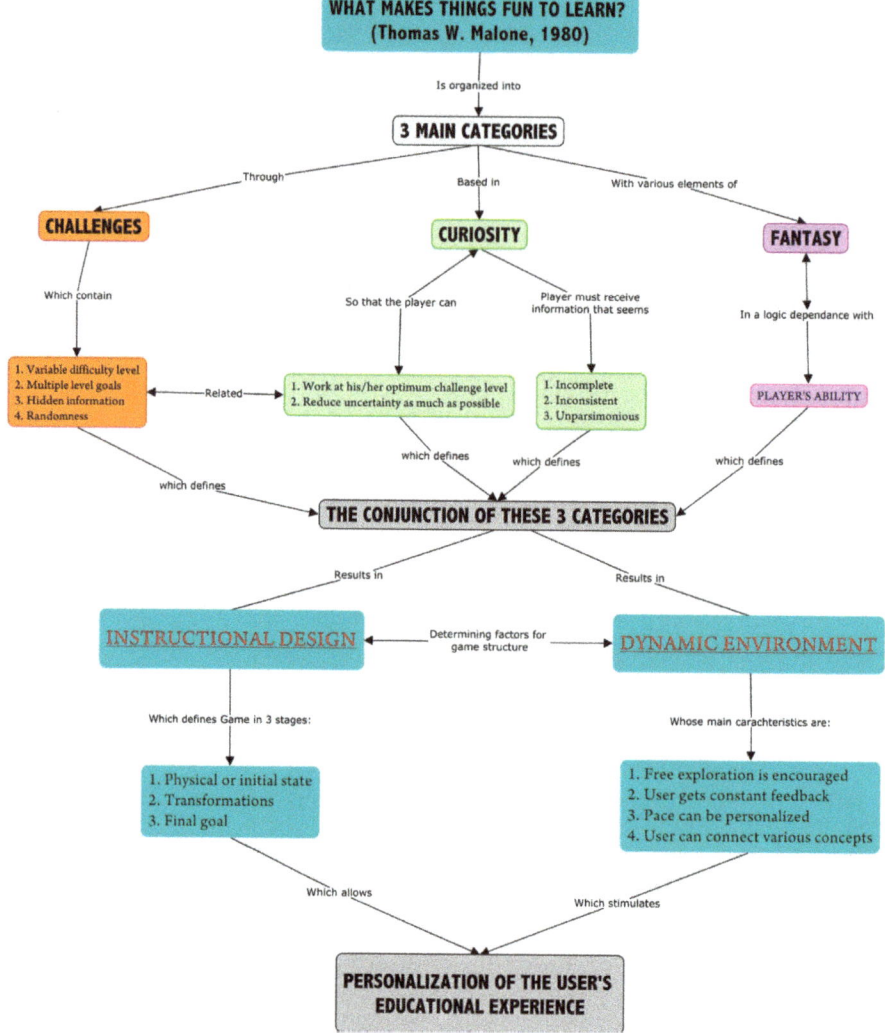

Figure 3. Conceptual map of Thomas W. Malone's Ph.D. dissertation (source: personal elaboration).

Performance levels will be evaluated based on quartiles, as it is the best mathematical way to understand the spread of a data set. Even when other alternatives could be considered (such as dividing the final grades into pass/not pass or dividing into quintiles), we consider this as the most realistic way to split student's qualification, as it clearly marks the distinction between the passing point (50%) and another "quality" measurement within the pass (75%) and not-pass (25%) grades, making it a very good tool to understand where each of the students lie within this evaluation ladder:

1) Q1 (bottom 25% of the total): This quartile will define the lowest performance level possible for a given variable. This quartile will be given to students that are substantially underperforming based on their capacity and the teacher's evaluation.
2) Q2 (bottom 50% of the total): This quartile will define the second lowest performance level possible for a given variable. This quartile will be given to students that are below performance. Students

falling under this quartile will be evaluated as fulfilling the most basic objectives for a specific variable, always based on the individual student's capacity and the teacher's evaluation.

3) Q3 (top 25% of the total): This quartile will define the second highest performance level possible for a given variable. This quartile will be given to students that perform satisfactorily or even slightly remarkably for a specific criterion.

4) Q4 (100% of the total): This quartile will define the highest performance level possible for a given variable. This quartile will be given to students that have an outstanding performance.

The criteria for evaluating active methodology experiences in class will now be defined. Five criteria were chosen, because this amount will most likely help us cover all crucial factors to be adapted from Malone's game theory and make the final grade easily calculable. To provide an example, the Faculty Innovation Center at The University of Texas at Austin [22] provides a guide to determining criteria to assess student work as follows:

a) Decide which of those criteria are "non-negotiable."
b) Ideally, your rubric will have three to five performance criteria.
c) What are the learning outcomes of this activity?
d) Which learning outcomes will be listed in the rubric?
e) Which skills are essential at competent or proficiency levels for the task or assignment to be complete?
f) How important is the overall completion of the task or project in comparison with other factors?

Thus, based on all of the above, it is necessary to define the five criteria that will be included in our evaluation (usually these activities will be performed by groups of students, so even when these criteria will be evaluated individually, they will also need to reflect the didactical approach to a group organization). Based on Malone's ideas (from Figure 3), each of his concepts will be redefined to match the educational goals of practicalizing gamification in a more didactical way. Therefore, the following criteria will be defined:

- *Challenge*: The concept of challenge is one of the most interestingly exposed by Malone. The author defines it as one of the main characteristics of intrinsically motivating environments for games [7]. He defines four main variables that define how challenge is presented to the user and achieved in game. When these criteria (difficulty level, multiple goals, hidden information, and randomness) are put together, they place a substantial weight on defining both settled goals and performance. This is the reason why, when translating this into a more academical/didactical approach, there will be a connection between goals, performances, and achievement perceptions, both from the teacher's and the student's perspective. By evaluating achievement, the student will be evaluated on the following:

    o how well the activity challenge responds to the student's idea of what an intellectual challenge will look and feel like;
    o how optimally the innovative activity is able to provide, within the group of students, different difficulty levels for the same challenge;
    o how the group will react to the challenge from an achievement perspective, given the fact that students will know how they will be evaluated prior to starting the activity in the classroom.

- *Curiosity:* Throughout his study, Malone uses many different references that are quite useful in defining the concept of curiosity that he uses. A good example related to games is the one from Ellis and Scholtz where they proved that "novelty was very important in determining which toys a child began playing with" [23]. Malone makes use of these ideas: "The kind of complexity or incongruity that is motivating is not simply a matter of increased information (...) Rather it involves surprisingness with respect to the knowledge and expectations a learner has." Based on these explanations, there is a clear relation between novelty, "surprisingness," and curiosity. For the sake of performing a holistic didactic evaluation in the rubric, the best way to assess these ideas in students is to use the concept of originality, as by stating this goal students will be evaluated (by the teacher and/or other students) on the following:

- how well the activity challenges students when firstly confronted with the idea to be developed, as the goal is to make an activity that will "oil the wheels" of their brains and that preserves the whole creation process as much as possible;
- how many surprise elements are embedded into the activity, the extent to which the teacher can add elements of incompleteness, inconsistency, and unparsimoniousity in class to make their students reflect about their work;
- how original (without losing a sense of quality for the final content) the final "product" presented by the group of students is and how each of them has contributed to it from an individual perspective.

- *Fantasy*: The author differentiates deeply between intrinsic and extrinsic fantasy. He writes: "In an extrinsic fantasy, the fantasy depends on the user's use of the skills but not vice versa. Most extrinsic fantasies depend on whether or not the skill is used correctly (i.e., whether the answer is right or wrong). (...) In intrinsic fantasies, on the other hand, not only does the fantasy depend on the skills, but the skill also depends on the fantasy. (...) In intrinsic fantasies, the event in the fantasy usually depends not just on whether the skill is used correctly, but on how its use is different from the correct use." [2]. Malone ends his dissertation recommending the use of intrinsic fantasies for games.

  Taking these ideas into the cognitive aspects of fantasy, Malone cites Petrie [24] claiming that "some kind of metaphor or analogy is epistemologically necessary for anyone to ever learn anything radically new because new knowledge can only be comprehended in terms of old knowledge." Malone continues with the most crucial part of his argumentation saying: "The final cognitive advantage of intrinsic fantasies is simply that, by provoking vivid images related to the material being learned, they can improve the memory of the material." When translating this into the objectives of this paper, it is clear that the author speaks about two specific terms: intrinsic fantasies (and their relationship with students' skills) and the necessity to use metaphors and analogies to build memory-based knowledge. The best way to connect both interesting ideas for the learning process (skills and memory) is through *motivation*. Motivation is the central idea that Malone wants to assess when writing his dissertation, and it is the quintessential idea related to fantasy. When exposed to an intrinsic fantasy (in the form of an activity) that will react to students' skills, they will be able to see that the level of skills required for them (as a group and as an individual) is adaptable to what they are supposed to be doing, and this will boost motivation within the group, as they will feel that they have the skills (or are about to acquire them) to solve the problem. Additionally, by provoking vivid images related to the material being learned, students are confronted with a problem that provokes an emotional reaction. This will allow the teacher to evaluate both their emotional and memorial response on their final assessment. Based on this motivation idea, students will be evaluated based on these criteria:

  - how intrinsic the reaction from the students is and what emotional response occurs in the class as a whole when presented with the idea of the activity;
  - how the proposed activity will boost motivation in the students by including all different kinds of game elements into it (such as vivid images, riddles, and metaphors);
  - how the activity's requirements will affect student's motivation.

- *Design*: After evaluating the three main categories that shape Malone's game theory (challenges, curiosity, and fantasy), he proposes that the conjunction of these three categories result in two criteria that determine the factors of the game structure: instructional design and a dynamic environment. The meaning of the first category was transformed into a more practical approach by calling it *organizational design & quality*. Students will be evaluated not only in the three criteria mentioned before but also in their resolution in relation to the presented final product. In other words, games are evaluated in their design by studying how different variables interact with one other, making the game more or less attractive, and the student's work during the activity will be evaluated similarly. It is critical that an element of quality and design is included into the rubric, as this is a more "traditional" method of evaluation, and it should also be part of this holistic, yet innovative proposal. Students, for this paragraph, will be evaluated on the following:

- how well the activity covers all objectives from a teacher's perspective;
- how well the students respond to the organizational challenge of writing, searching for information, and preparing to present work in front of the teacher and the rest of the students;
- how many high quality standards (specified for each and every activity) students have been able to fulfill during the planning and presentation phase.

- *Environment*: The second determining factor for Malone is a dynamic environment. Malone defines how the environment should respond to the user's activity in many ways, but one of the most relevant for this study is that "an environment should be both responsive to the learner's activities and helpful in letting him take a reflexive view of himself." [7]. This, in relation with a more pedagogical approach to game theory and taking into consideration all possible variables related to the actual consecution of objectives and goals for the students, could be defined as the fifth and last criterion: *relationships & time management*. Environment, as it will be understood in this study, refers to surroundings that affect the individual student when working on a specific activity. As these activities students are often settled upon different groups, it is imperative to include a "social factor" within the rubric, to ensure this determining angle is not left apart. Time management is also an extremely interconnected factor with relationships between the members of a group when performing together, so it is also included for the sake of generating a full-holistic evaluation. For this specific criterion, students will be evaluated on the following:

    - how smoothly the activity is conducted during the definition and execution phase, and how well the activity was shared and presented in front of the other students;
    - how "on time" the activity is presented and whether students have respected all deadlines for the activity;
    - during presentation in class, how respected the time given to them as a group is and how well individual responsibilities are met (who is the leader or acts like one?)

All criteria have been defined for the evaluation rubric, so here we define it as follows on Table 1. Some common descriptive terms to indicate a progression from the Faculty Innovation Center at The University of Texas at Austin [22] have been used to model this rubric.

In the example shown above, the theoretical student would achieve 80% in this activity; out of 20 possible points, he/she would achieved 16. Students must know before starting any activity on which aspects will they be evaluated and how this evaluation will affect the rest of their unit's evaluations. By using this tool, teachers will be able to use game theory ideas to evaluate from a comprehensive point of view, all factors that will be taken into consideration when proposing, from a teacher's point of view, any kind of activity related to gamification and other active methodologies.

Table 1. An example of a qualitative proposal for an individual evaluation rubric based on Malone's corollaries (source: personal elaboration).

Criteria	Q4 (4/4)	Q3 (3/4)	Q2 (2/4)	Q1 (1/4)	Evaluation
Achievement (20%)	All activity tasks have been fulfilled and presentation was fully developed.	Most activity tasks have been fulfilled and presentation was mostly developed.	Some activity tasks have been fulfilled and presentation was partially developed.	Few activity tasks have been fulfilled and presentation was not developed.	4
Originality (20%)	Broad, highly varied and non-repetitive concepts for this activity have been presented.	Adequately varied but occasionally repetitive concepts have appeared.	Quite limited document with lack of variety and repetition of ideas.	Very limited, basic, presentation has been memorized and highly repetitive.	3
Motivation (20%)	Always show desire to work. Skills related to the activity have all improved.	Usually show desire to work. Skills related to the activity have mostly improved.	Some of the time show desire to work. Skills related to the activity have improved to some extent.	Rarely or not at all show desire to work. Skills related to the activity have not improved.	4
Design & Quality (20%)	Fully developed document, supported by a very high-quality standard.	Adequately developed document, adequately supported by an ok-quality standard.	Partially developed document supported with low-quality standard.	Minimally developed document supported with a very low-quality standard.	3
Relationships & Time Management (20%)	Working together, interacting, improving and always on time.	Most of the time working together, interacting and almost always on time.	Main group working together, not everybody is involved and poor time management.	Isolated individuals, no interaction and awful time management.	2
				RUBRIC GRADE	4+3+4+3+2 = 16/20
				FINAL GRADE	80%

4. Discussion—Expected Results

Based on the working hypothesis, and always taking into consideration the goal to assess a qualitative initiative from a theoretical perspective, the results of designing this rubric could provide a very well-structured answer to the debate of ensuring that a proper evaluation is performed when deciding to work with any kind of innovative methodology.

It is necessary to highlight that, apart from proposing evaluation percentages for each of the rubric's criteria, it is crucial that, based on the future results of this proposal, how the didactic unit percentages will reflect these activities' weight in the total weight of the evaluation process can be evaluated.

The evaluation process will be composed of several percentages for all of the classroom and homework's activities proposed to the students—it will be a mix of rubrics and other evaluation materials or even a mix of different rubrics. Based on this idea, some authors have proposed in previous studies [25,26] various alternatives for the final share of this percentage sharing. Based on the idea embedded throughout this paper, defending the idea that the final evaluation of the student must be as holistic as possible and must entitle students to be positively evaluated if they can prove to be sufficiently prepared in different evaluation aspects such as exams, homework, classwork, and activities of active methodologies, the teacher's evaluation method must include all of these different projects to ensure that proper objective evaluation is performed.

As a matter of exampling, the percentage sharing exposed here will be based on our own teaching experience and evaluation system. Considering 100% as the total available grades for a specific unit, students will be evaluated as follows:

A. **35% based on the rubric:** Out of the 100% final grade for a specific unit, 35% will be based on the final grade obtained in the rubric.
B. **35% based on exams:** This percentage will include all questions on the partial exams (if applicable) and the final exam. These exams must include various different questions (test-type,

explanatory, etc.) so that all possible "classic" didactic criteria will be assessed by answering these questions.

C. **30% based on classroom performance and homework**: All classroom and homework activities (taking notes, reviewing exercises, class participation, volunteering, etc.) must be considered in the evaluation and must be weighted equally with the rest of the activities that students will perform throughout the unit.

The implications of this holistic approach will be considered in the broadest context possible, especially considering that 65% of the final evaluation of all the students will be based on their day-to-day classroom/home performance and their ability and knowledge expressed and evaluated through the rubric. Additionally, by proposing this "equally-weighted" evaluation for these three criteria, it is expected that all actors involved in the educational process will understand the message that teachers are sending: grading must be a matter of evaluating consistency instead of just giving value to remembering what you have been instructed to write down on the exam day. This way it could be ensured that, as an expected result, educational grading systems will start to reflect more accurately the real effort that students are putting into the learning process.

Based on all of the above, and for the sake of discussing the research questions after presenting the method used in this investigation, the main questions of this paper could be answered as follows:

- Could learning be improved by using gamification strategies?

    It is clear that the proposed method can only be improved when measurements are applied to it. This is the reason why ideas such as achievement, originality, motivation, design, and quality must be evaluated in a structured way that will allow teachers and facilitators to improve them throughout the student's school cycle.

- If so, how will this be reflected in students' performance?

    Students' performance will be strongly affected by this new proposed method of evaluation, as up to 65% of their performance will depend on activities that will reflect their improvement on the above-mentioned concepts. This will allow students with a good attitude and a will to learn but with a bad performance in the "one opportunity" approach to perform better. Many of these students, for various reasons, might not perform well on the "100% exam" approach, but they might do better if 65% of their grades are based on a holistic approach.

- What is the cost of this learning? Can teachers take advantage of a new proposed evaluation method?

    The most important factor to take into consideration when proposing these activities to teachers is to assure them that, by "moving" into this new proposal, their workload not only will not increase but will eventually be reduced. Thus, in this regard, the cost is very low. By using this approach with rubrics, teachers can reduce evaluation time, as they can evaluate *while* students are presenting their work. Teachers could even store these tools virtually in any kind of device, making the evaluation process much smoother and available in multiple platforms. This is the main reason why teachers will take advantage of these new proposals.

5. Conclusions

The main conclusions of this work are aligned with student's learning enjoyment, the idea of *conceptual valuable enjoyment*, and the creation of an interactive framework to achieve improvements in the teaching–learning process.

As discussed throughout this article, gamification greatly enhances both cooperative and collaborative learning. Both the teacher and the student must change the role they play so that the student becomes the real protagonist of their own learning and the teacher plays a role of the facilitator of this process. By abandoning the traditional framework of the teacher being exclusively responsible for disseminating content to become a guide or reference, which is able to help the student to find its own way, teachers will help to develop a much more enjoyable and profound learning process for his/her students.

In the same way, gamification processes must facilitate student's interactivity with their peers, as in many situations they need to cooperate between them to find a solution to various proposed problems. This, at the same time, aligns very well with other relevant learning initiatives, such as problem-based learning and project-based learning, which opens a space of collaboration between heterogeneously interactive active methodologies.

Based on all of the above, two ideas will be considered basic and transversal to this proposal:

- Humans enjoy learning in general and are able to learn while enjoying.
- The final main idea is to take advantage and recognize the potential for gamification to be implemented in a framework that allows, in many cases, to rethink the game.

The theoretical and practical impact of this study could be best understood when taking into consideration the junction between these two transversal ideas and the four main research questions expressed and discussed throughout the paper. The implications of turning gamification theory into a practical tool (evaluation rubric) that could be used in many different evaluation activities could be diverse, especially thanks to its cross-cutting approach. It is expected that this impact will be better measured when implementing these tools in a real secondary class environment.

This is the reason why, as a proposal for potential prospective studies, it would be very interesting to investigate how the use of gamification improves motivation in students from a practical point of view. Additionally, it would be significant to highlight the possibility of developing this rubric based not only on Malone´s corollaries but also on other interesting authors such as Deterding et al. [11]. Furthermore, it could also be thought-provoking to propose other different frameworks that would put at teacher's disposal more resources (i.e., visual and interactive models) that might better adapt to different learning styles. The next learning paradigm will be in connection with these ideas.

Author Contributions: Conceptualization: D.C.M. and J.J.R.G.; formal analysis: D.C.M.; investigation: D.C.M.; methodology: D.C.M. and J.J.R.G.; project administration: D.C.M.; resources: J.J.R.G.; supervision: J.J.R.G.; validation: J.J.R.G.; visualization: J.J.R.G.; writing—original draft: D.C.M. and J.J.R.G.; writing—review & editing: D.C.M.

Funding: This research received no external funding.

Conflicts of Interest: The authors declare no conflict of interest.

References

1. Gamificación en el cine: Jumanji. Available online: https://www.wonnova.com/blog/gamificacion-cine-jumanji-201402 (accessed on 28 January 2019).
2. Calderón, Q.; Isabel, R. Diseño y Validación de una E-Rúbrica para la Evaluación de Competencias Clínicas Transversales de Bioética en Pediatría. Available online: http://dspace.casagrande.edu.ec:8080/handle/ucasagrande/1375 (accessed on 28 January 2019).
3. Masmitjà, J.A. Rúbricas para la evaluación de competencias. Available online: https://www.researchgate.net/publication/299903426_Rubrica_para_la_Evaluacion_de_la_Competencia_Innovacion_Creatividad_y_Emprendimiento_en_master (accessed on 28 January 2019).
4. Munuera Gómez, P.; Ruiz González, R. Gamificación, portafolio digital, contrato académico y rúbrica. estrategias para la adquisición de competencias. Available online: https://www.researchgate.net/publication/318827885_GAMIFICACION_PORTAFOLIO_DIGITAL_CONTRATO_ACADEMICO_Y_RUBRICA_ESTRATEGIAS_PARA_LA_ADQUISICION_DE_COMPETENCIAS (accessed on 28 January 2019).
5. Borrás Gené, O. Fundamentos de Gamificación. Monografía (Documentation). Rectorado (UPM), Madrid. Available online: http://oa.upm.es/35517/1/fundamentos%20de%20la%20gamificacion_v1_1.pdf (accessed on 28 January 2019).
6. Google Trends. Available online: https://trends.google.com/trends/?geo=US (accessed on 28 January 2019).
7. Malone, T. What Makes Things Fun to Learn? Heuristics for Designing Instructional Computer Games. Available online: https://hcs64.com/files/tm%20study%20144.pdf (accessed on 28 January 2019).

8. Pelling, N. The (Short) Prehistory of 'Gamification' Funding Startups (& Other Impossibilities). Available online: https://nanodome.wordpress.com/2011/08/09/the-short-prehistory-of-gamification/ (accessed on 28 January 2019).
9. Németh, T. English Knight: Gamifying the EFL Classroom. (Unpublished Master's Thesis), Pázmány Péter Katolikus Egyetem Bölcsészet- és Társadalomtudományi Kar, Piliscsaba, Hungary. Available online: https://ludus.hu/gamification/ (accessed on 28 January 2019).
10. Deterding, S.; Dixon, D.; Khaled, R.; Nacke, L. From Game Design Elements to Gamefulness: Defining "Gamification". Available online: http://www.rolandhubscher.org/courses/hf765/readings/Deterding_2011.pdf (accessed on 28 January 2019).
11. Deterding, S.; Dixon, D.; Khaled, R.; Nacke, L.E. Gamification: Toward a Definition. Available online: http://gamification-research.org/wp-content/uploads/2011/04/02-Deterding-Khaled-Nacke-Dixon.pdf (accessed on 28 January 2019).
12. Lucassen, G.; Jansen, S. Gamification in Consumer Marketing—Future or Fallacy? *Procedia Soc. Behav. Sci.* **2014**, *148*, 194–202. [CrossRef]
13. Gamification Co. Gartner Adds Gamification to its Hype Cycle. Available online: http://www.gamification.co/2011/08/12/gartner-adds-gamification-to-its-hype-cycle/ (accessed on 28 January 2019).
14. Gartner. Newsroom. Available online: https://www.gartner.com/en/newsroom (accessed on 28 January 2019).
15. Dicheva, D.; Dichev, C.; Agre, G.; Angelova, G. Gamification in Education: A Systematic Mapping Study. *Educat. Technol. Soci.* **2013**, *18*, 75–88.
16. Khan Academy. Available online: http://www.khanacademy.org (accessed on 28 January 2019).
17. Witt, M. Gamification of Online Idea Competitions: Insights from an Explorative Case. Available online: https://www.researchgate.net/publication/267365902_Gamification_of_Online_Idea_Competitions_Insights_from_an_Explorative_Case (accessed on 28 January 2019).
18. Khatib, F.; Cooper, S.; Tyka, M.D.; Xu, K.; Makedon, I.; Popović, Z.; Baker, D.; Players, F. Algorithm discovery by protein folding game players. *PNAS* **2011**, *108*, 18949–18953. [CrossRef] [PubMed]
19. O'Donohoe, S.; Vedrashko, I. Game-Based Marketing: Inspire Customer Loyalty Through Rewards, Challenges, and Contests. *Int. J. Advert.* **2011**, *30*, 189–190. [CrossRef]
20. Hamari, J.; Koivisto, J.; Sarsa, H. Does Gamification Work? A Literature Review of Empirical Studies on Gamification. In *Proceedings of 47th Hawaii International Conference on System Sciences, Waikoloa, HI, USA, 6–9 January 2014*; IEEE Computer Society: Washington, DC, USA.
21. Popham, W.J. What's Wrong—and What's Right—with Rubrics—Educational Leadership. Available online: http://www.ascd.org/publications/educational-leadership/oct97/vol55/num02/What\T1\textquoterights-Wrong%E2%80%94and-What\T1\textquoterights-Right%E2%80%94with-Rubrics.aspx (accessed on 28 January 2019).
22. The University of Texas at Austin. What is a Rubric? Available online: https://facultyinnovate.utexas.edu/sites/default/files/build-rubric.pdf (accessed on 28 January 2019).
23. Ellis, M.J.; Scholtz, G.J.L. *Activity and play of children*; Prentice-Hall: Englewood Cliffs, NJ, USA, 1978.
24. Petrie, H.G.; Oshlag, R.S. Metaphor and Learning. In *Metaphor and Thought*, 2nd ed.; Ortony, A., Ed.; Cambridge University Press: Cambridge, UK, 1993; pp. 579–609.
25. Martin, C.; Horton, M.L.; Tarr, S.J. Building Assessment Tools Aligned with Grade-level Outcomes. *J. Physical. Educ. Recre. Dance* **2015**, *86*, 28–34. [CrossRef]
26. MIT Teaching and Learning Laboratory. Grading Rubrics. Available online: http://tll.mit.edu/help/grading-rubrics (accessed on 28 January 2019).

© 2019 by the authors. Licensee MDPI, Basel, Switzerland. This article is an open access article distributed under the terms and conditions of the Creative Commons Attribution (CC BY) license (http://creativecommons.org/licenses/by/4.0/).

Article

Teaching HCI Skills in Higher Education through Game Design: A Study of Students' Perceptions

Pedro C. Santana-Mancilla [1,*], Miguel A. Rodriguez-Ortiz [1], Miguel A. Garcia-Ruiz [2], Laura S. Gaytan-Lugo [3], Silvia B. Fajardo-Flores [1] and Juan Contreras-Castillo [1]

1. School of Telematics, University of Colima, 28040 Colima, Mexico; maro@ucol.mx (M.A.R.-O.); medusa@ucol.mx (S.B.F.-F.); juancont@ucol.mx (J.C.-C.)
2. Department of Mathematics and Computer Science, Algoma University, Sault Ste. Marie, ON P6A2G4, Canada; miguel.garcia@algomau.ca
3. School of Mechanical and Electrical Engineering, University of Colima, 28400 Coquimatlan, Mexico; laura@ucol.mx
* Correspondence: psantana@ucol.mx

Received: 22 March 2019; Accepted: 10 May 2019; Published: 14 May 2019

Abstract: Human-computer interaction (HCI) is an area with a wide range of concepts and knowledge. Therefore, a need to innovate in the teaching-learning processes to achieve an effective education arises. This article describes a proposal for teaching HCI through the development of projects that allow students to acquire higher education competencies through the design and evaluation of computer games. Finally, an empirical validation (questionnaires and case study) with 40 undergraduate students (studying their fifth semester of software engineering) was applied at the end of the semester. The results indicated that this teaching method provides the students with the HCI skills (psychology of everyday things, involving users, task-centered system design, models of human behavior, creativity and metaphors, and graphical screen design) and, more importantly, they have a positive perception on the efficacy of the use of videogame design in a higher education course.

Keywords: learning by doing; serious games; game design; human computer-interaction; HCI education

1. Introduction

Human-computer interaction (HCI) courses often use the curriculum of the Association for Computing Machinery Special Interest Group on Computer-Human Interaction (ACM SIGCHI) as a foundation [1]. As reported by Greenberg [2], this curriculum should not be used directly when it comes to an individual course within a career, rather as a complete curriculum to train HCI specialists.

In the software engineering program in the School of Telematics at the University of Colima, HCI is an individual course taught in a project-based approach to engage students in research of real problems [3], thus providing students with the basic knowledge of HCI and the different areas that converge in this field is ideal. Computer games (also called video games) are a rich opportunity to bring this approach into a traditional HCI course.

In this paper we present two main contributions: first, we present a proposal for using the design and evaluation of computer games as a learning tool to teach HCI to undergraduate students, and, second, we provide an empirical validation of this proposal to explore students' attitudes to understand if the students believe that using video games allow them to learn higher education skills. The objective of this research was to explore HCI students' perceptions regarding the use of game design to acquire the needed skills in the subject, and the research question that we aimed to answer was: does the use of videogame design improve the students' perception of their learning?

2. Teaching Human-Computer Interaction

A report published by Hewett et al. [1], in addition to giving a formal definition of human-computer interaction, proposed a syllabus and contents for HCI courses in which students could understand the multidisciplinary nature of the field, leaving an open range of topics that interact in the academic training in HCI.

Many years have passed since the publication of that report and the research in the area has grown a lot. However, until recent years, the conclusions were the same. As Churchill et al. [4] states, there is still no agreement on the range of topics that integrate the area and on how to teach HCI courses, how much fundamental theory must be presented, and how much practice students should perform; there are divided opinions about it, and they are investigating the philosophies and best practices to update the ACM 1992 curriculum proposal and support the present and future of HCI education [1].

Despite the lack of agreement on the percentages of theory versus practice, there is a point where most converge since that first report [1]: the development of human-computer interfaces is a matter of engineering and design, and, the subsequent learning of theoretical content, is enriched with the cooperative experience and can advance to a more mature level when students are required to solve the problems of real-life projects. Hartfield et al. [5] also mentioned that a combination of readings, classes, and written assignments is appropriate for teaching topics such as solutions to typical design problems and theories of human skills that are related to HCI; however, they also considered that these activities do not help HCI students to acquire practical skills in the design of technological solutions, and recommended hands-on exercises be carried out with real projects so students can learn by doing, as suggested by Pastel [6]. His work also mentioned that learning through real projects in collaborative groups is becoming the standard for the teaching of the HCI, and he made the observation that the students prefer this method of teaching since they can learn by doing and carrying out a meaningful project with their projects designs. Pastel [6] also made an essential contribution by adding research techniques to work, this was because, as he mentioned, there is a significant similarity between the scientific research process and the software development process. In addition to being research projects, students can use unique HCI devices or implementations.

Lorés et al., Koppelman et al., Reimer et al., Solano, and Urquiza-Fuentes et al. [7–11] also used a combination of theory and projects with real-life applications using user-centered design methodologies. In fact, Lorés et al. [7] combined the teaching of formal content with the development of a real-life project in a compulsory course within the HCI curricula at the University of Lleida (Spain). The work of Koppelman et al. [8] consisted of providing projects with a realistic context, inviting people from industry to serve as clients for the student projects and allowing teams to evaluate their designs with real users. Reimer et al. [9] used the studio-based approach to teach competencies in designing interactive objects in an HCI design studio course. Solano [10] presented an analysis of the teaching experience in an HCI course at the Universidad Autónoma de Occidente of Colombia and identified that problem-based learning is a good way to construct knowledge in real scenarios that can be improved. Urquiza-Fuentes et al. [11] presented a study to test the effect of using practical exercises in a human-computer interaction course, and they found that realistic projects are a viable approach to teach HCI and that students involved in this approach are significantly more motivated.

On the other hand, as mentioned by Serrano-Cámara et al. [12], it is essential to keep the students motivated, especially in generating intrinsic motivation [13] (which refers to doing something because it is inherently interesting or pleasant) because motivation is a central factor in learning and, as mentioned by the authors, improves results in knowledge and creativity. Urquiza-Fuentes et al. [11] suggested that collaborative learning encourages the use of high-level cognitive strategies, critical thinking, deep learning, deep understanding, and positive attitudes towards education and teamwork, according to the research that real-life project development combined with collaborative learning approaches keeps the students motivated; therefore, it turns out to be a positive strategy for teaching the HCI.

3. Learning with Games

The last report from the Entertainment Software Association [14], showed that the videogame industry has maintained a steady growth and a prominent place in the market, being one of the most successful application domains in the history of interactive systems [15].

Video games have not only been used for entertainment but also as an educational tool since the design of games for education provides valuable help in skills acquisition [16]. The design of video games with and educational purpose is known as serious games, a relatively new discipline that combines learning design with game mechanics and logic [17].

Williamson et al. [18] argued that video games are a compelling context for learning because it is possible to create and interact with virtual worlds to develop situated and contextualized understandings. Robertson and Howells [19] mentioned that there is research that argue to give players a mental exercise that generates a series of cognitive abilities, such as the planning of strategies for the resolution of problems, and also mentioned that not only the use of games contributes to learning, rather, they suggested the benefits of designing and building games as part of learning strategies since it involves students and education, and thus becomes an active experience.

In addition, Skrzyszewski et al. [20] remarked on the benefits of including the design and construction of video games as a learning strategy: this activity creates the opportunity to develop and increase creativity, technical skills, and the ability to work as a team. In addition, students can acquire an update on state-of-the-art technologies and techniques used in the process, and gain practical experience in different areas of computing. Although there exist arguments against learning with games, centered upon the lack of empirical evidence to support their effectiveness [17], we found that according to [21–23], there is pedagogical support for learning through the development of video games and that, in addition to the aforementioned, it increases motivation and makes the courses more attractive for students.

There are also examples of the application of video game design and development activities as part of HCI courses, such as the case of Dyck et al. [15] who mentioned that these environments do not put restrictions on how things should look or how the interaction should be, but reward innovation and performance that often leads to novel interaction, which have led to the production of many ideas for future work. Bernhaupt et al. [24] agreed that the design of video games, as part of an HCI course, allows for exploring new interaction techniques and experiences, allowing faster adoption of novel technologies, and, as an example, they mention the creation of Oculus Rift and the Kinect, of which Villaromen et al. [25] have proposed the latter as projects for HCI courses.

In addition, as mentioned by Covaci et al. [26], modern educational theories describe effective learning as active, experiential, situated, problem-based, and providing immediate feedback. The development of games can be the balance between play and learning activities, including this type of events in HCI courses, which is consistent with the theories of effective learning since, with the design and development of games and real-world situations, that allows students to develop skills for solving problems in a way that they also stay motivated [27,28].

4. Experimental Study

This study was implemented in a fifth semester mandatory human-computer interaction course, taught at the University of Colima in Mexico for the software engineering major. This course has been taught since 2009 and, during these years, the learning has been mixed, using projects with video games and learning with utility projects or administrative software.

The experiment was between subjects [29], in which students were taking the course when video games were used. This evaluation approach is a study design in which each group is only exposed to a single setup (i.e., learning with utility software or learning with video games). At the time of the study, the student population of the fifth semester in software engineering were 40:29 (72.5%) were male, 11 (27.5%) were female, and none of them had taken an HCI course before. In this stage of their education, they already had skills in programming, data structures, and software development processes.

The study was designed to explore students' perceptions of the efficacy of video game design to acquire the required HCI competencies: knowing the importance of the correct design, application and evaluation of human computer interfaces, identifying paradigms, and state-of-the-art in the design of technological developments with the purpose of producing efficient, easy to use software interfaces to solve computing problems.

5. The Games

In this section we describe two examples of computer game projects that were designed and developed during the course. These games were selected because they were success cases since the students published their results in research conferences [30,31].

5.1. Fallbox: Game Controlled by Head Tracking

Fallbox is a 2D game (see Figure 1) played from the first-person perspective. It is based on head tracking using infrared led and the infrared camera of the Nintendo Wii Remote. The aim of this game is that players move their heads to control the game avatar (called Chilo) on the X axis. Chilo has to move to dodge the boxes that are falling down.

Figure 1. The computer game Fallbox. The avatar must dodge the falling boxes.

The game has 5 levels, with increasing difficulty in each level, because the boxes are falling faster. The player has a limited life (a 100% bar at start) that decreases each time Chilo is hit by a falling box.

5.2. Shooter: Game with Multimodal Interaction

This game is a 2D with a first-person perspective. The characteristics and goals of the game are simple: the user has to go through different scenarios in which he has to eliminate every enemy that appears within a time frame. The game integrated three different types of tangible multimodal interactions (see Figure 2):

1. Pen-based interface: these kind of interfaces uses stylus-type devices to point to elements on the user interface or to simulate handwriting [32];
2. Vision recognition: this recognition technique is used to simulate the mouse by wearing colored tapes on the user's fingers for tracking the movements and controlling the cursor [33]; and
3. Gesture interface: this interaction is achieved using a video camera and computer vision techniques to capture the hand shape and movement, in order to simulate the mouse motion [34].

Figure 2. The students create the interacting devices. The image illustrates a gesture interaction.

6. HCI Teaching Process

The course has been oriented to teaching through the design and evaluation of computer games and their interaction devices. This way students learn not only basic game design principles, but also acquire skills from the different areas that converge in the HCI area.

The course follows the proposal of Greenberg [2] that presented the teaching of HCI as a usability engineering process (see Figure 3), which has the purpose for students to acquire sufficient skills to make a reasonable design, develop, and evaluate of human-computer interfaces. Upon completion of the course, students must understand what a good design means, and have experience designing systems that are usable for people, which must be achieved by implementing interfaces through prototypes and practicing methods of evaluating the quality of the product.

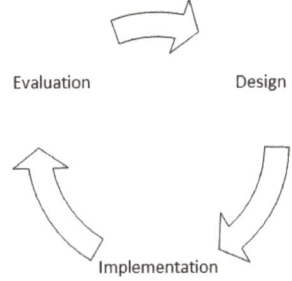

Figure 3. Human-computer interaction (HCI) as a usability engineering process.

6.1. Process of Design

The design process in HCI is fundamental and designing includes learning and acquiring important knowledge by the students [2]: psychology of everyday things, involving users, task-centered system design, models of human behavior, creativity and metaphors, and graphical screen design.

The design of a game starts with a plan. It will help define the game story and its elements. The game plan contains the designs of characters and scenarios, in addition to the types and levels of interaction that the player will have with the game. To achieve this, the sketching technique proposed by Buxton [35] was used. This technique has proved to be an excellent means for people to start thinking about the process of product design in a simple way [36].

To evaluate the students' success during this stage of the course, students must generate the sketches with the initial design of the game and the interaction mechanisms (for examples, see Figures 4 and 5).

Figure 4. Design of the game mechanism.

Figure 5. Design of the interaction with the game control device.

6.2. Process of Implementation

The emphasis of the development process is on the behavioral domain: the tasks are concerned with the user interaction with the application and the view of the user. A variation of the user-centered methodology is used based on user-task analysis, where task-oriented methodologies ask that users attempt to complete a task in the system [37]. The first stage in the process is to identify a complete description of tasks, subtasks, and methods required to use a system, in addition to having a complete understanding of the user. With this process, designers have a clear understanding of the user requirements, and, therefore, could design better functionality according to the user. After gathering the users' task analysis, the constructional domain starts, which involves modelling, source code generation, and integration.

For the Fallbox game, in this stage the students identified the scenario of the game as a complete description of the user task:

> "Chilo is the youngest kid in the family. For this reason, his brothers throw boxes to stop him inside their house. The player must help Chilo avoid these boxes by moving his head and show his brothers that he can beat them."

Once the tasks were defined, the subtasks were defined: the avatar will interact only with falling boxes, the boxes will fall with increased speed, the scenarios will include non-playable objects (i.e., tables, chairs, beds, frets, stove, refrigerator), and the avatar will be controlled by the users' head movements.

At this point, the students have a clear understanding of the requirements, and they can start to build the game and the interaction devices. For the aforementioned game, they envisioned a device that should be placed on the player's head for head tracking that consisted of an infrared LED (see Figure 6) that points to a Wii Remote and is the responsible for indicating the position of the player to the avatar in the game. So, the player has to move her head to control the game and dodge the boxes that are falling down.

Figure 6. Head-tracking interaction device.

6.3. Process of Evaluation

To learn the competencies for the usability assessment, the students used a summative evaluation following the methodology proposed by Santana-Mancilla [38], called "IHCLab Usability Test for Serious Games".

The evaluation included the followings phases:

1. The moderator (course instructor) opens the session with an introductory text and applies the questionnaire for user characterization;
2. Each team gives a live demo of the game they developed in the course. The aim of this is to put into context the use of the controllers for the users;
3. The participants should be given a task list to complete on the game. Each team will evaluate the usability of the games developed by the other teams; and
4. The players answer the questionnaires to collect their opinions:

    (1) Game heuristics questionnaire: an adapted and generalized game heuristics instrument based on previous applications reported by the literature [39,40].
    (2) Game experience questionnaire (GEQ): The GEQ is divided in two dimensions: (1) four questions, where the learners had to give a score from 1 to 10, where 10 is the most significant, and (2) seven questions that measured some important indicators with a 5-point Likert scale.

7. Case Study

To validate the effectiveness of computer games use in the learning of HCI, as mentioned in Section 4, a between-subjects study was conducted with the two groups of students who followed the HCI course at the School of Telematics of the University of Colima in Mexico.

7.1. Subjects

Participants were 40 undergraduates studying their third year (fifth semester) with a software developer background (programming, algorithms, and data structures) and a basic hardware knowledge background (embedded systems and digital electronics). The school gave the informed consent for inclusion before the students were included in the study.

7.2. Questionnaire

A questionnaire (designed by our research group) to know the opinion of the students about their perception of the use of video games in HCI teaching was applied.

To validate the questionnaire contents, a group of experts in the area of HCI (expert judgment) used item analysis. The validity of content is essential when making inferences or generalizations from the results of the evaluation [41].

For this evaluation, four criteria were established:

- Sufficiency: The items that measure an indicator are enough to obtain the measurement of it;
- Clarity: The item is easily understood, that is, its syntactic and semantics are adequate;
- Coherence: The item has a logical relationship with the indicator that it is measuring; and
- Relevance: The item is essential or important, that is, it must be included.

To evaluate each criterion, the following response options were established:

1. Does not comply.
2. Low level.
3. Moderate level.
4. High level.

A total of five teachers, working in different HCI areas, validated the questionnaire. From the evaluation, a set of changes were introduced in the structure of the questionnaire (language used, closing items, rearrangement of some questions), developing the version that was used in this study, which consists of six questions and is shown in Appendix A. It is important to mention that this questionnaire had already been applied in other studies by our research team [42], but is still in the process of further validation using the Cronbach's Alpha coefficient to validate the internal consistency of the instrument since it is calculated from the covariance between the elements analyzed. The results obtained in this study will contribute to the validation process of the questionnaire.

Each of the students in the intervention groups filled out the questionnaire and, since all students accomplished the instrument, the sample includes all the perceptions of the subjects under study and, thus, meets the requirement for saturation.

7.3. Results

95% of the students participating in the course thought that they learned the required HCI skills using the design and development of video games (see Figure 7). The remaining 5% considered that it was necessary to integrate additional content so that learning was more complete. The topics most commonly proposed for improvement were evaluation of video games, novel interaction devices, and graphic design.

This result showed a largely positive perception of the efficacy of use video games for skills development in higher education.

When asked how they thought their learning would have been if the teaching process were not with video game design, 100% said that the knowledge acquired would have been lower. Very related, the 85% percent mentioned that they would attend other courses that use video games as a learning tool.

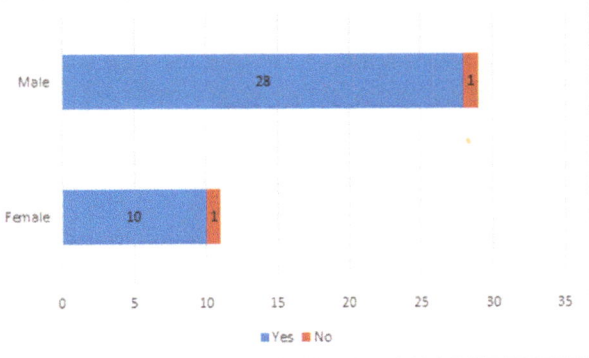

Figure 7. Number of students' responses about acquiring the required HCI skills by gender.

Regarding if they enjoyed learning using computer games, 100% of the students enjoyed the course, and the following is some representative feedback:

- *"I enjoyed from the moment of creating the idea of what game to develop, the design, the development of the new interaction and playing the new videogame."*
- *"The implementation of high technology interfaces and the evaluation to improve our project."*
- *"The use of videogames to compare the course topics with experiences closer to the common and everyday things, the work with devising a new form of interaction for a game was good and challenging."*

Regarding the problems, difficulties or frustrations they could have during the project, they said the following:

- *"On some occasions it was difficult for me to understand how interaction will work."*
- *"Understand the interaction that was going to be implemented and how it would be applied."*
- *"Actually, the biggest difficulty is learning to develop this type of applications, but as is more fun we were more motivated to learn."*

Finally, the students made some suggestions to improve the experience.

- *"I would suggest that projects like this used on the HCI course, be proposed as integrative projects for all the courses in the semester, because these projects motivate us to be involved deeper with the technologies implied."*
- *"I think that this way of learning is very good."*
- *"I would like there to be more time for the implementation of the interactions."*

8. Conclusions

Forty students who had participated in the HCI course were interviewed with an instrument following the conclusion of the semester. Participants were drawn from the human-computer interaction course of their undergraduate degree.

The use of computer games applied to HCI education has given us, in addition to learning, a motivating way of transmitting knowledge to students. The findings contribute to the learning-by-games literature by showing that students perceived value in learning by game design at the university level.

The approach of learning HCI through the design and evaluation of computer games has given results outside the course. To date we have nine undergraduate students who have created a computer game as a final project of their engineering studies based on the HCI course they attended. In addition, students and professors have published more than 15 research works (book chapters, journals, and conferences) in the field of video games and serious games with these projects and the lessons learned from the HCI course in the years that it has been offered.

By its nature, games offer constant challenges that students must be overcome to acquire the required skills, and these challenges allow the students to perceive that games can improve their required skills.

The evaluation has shown a broadly positive student attitude on how this game approach was perceived, demonstrating that, according to the perception of the students, games can have a positive effect on higher education.

The students' perceptions described here appear to require further investigation. These results have motivated a long-term randomized controlled trial where participants will be assigned to either an intervention or a control group to offer compelling evidence for the potential of designing video games to aid in the development of higher education competencies on HCI.

Author Contributions: Conceptualization, P.C.S.-M. and M.A.R.-O.; Formal analysis, P.C.S.-M., M.A.R.-O., M.A.G.-R., L.S.G.-L., S.B.F.-F., and J.C.-C.; Investigation, P.C.S.-M.; Writing—original draft, P.C.S.-M. and M.A.R.-O.; Writing—review & editing, M.A.G.-R., L.S.G.-L., S.B.F.-F. and J.C.-C.

Funding: This research received no external funding.

Acknowledgments: We thank all the students who participated in the case study during their HCI course.

Conflicts of Interest: The authors declare no conflict of interest.

Appendix A

(1) Do you think you learned enough about HCI through the design and evaluation of a videogame?
( ) Yes ( ) No
(2) Do you think your knowledge of HCI would have been the same, smaller, or greater without having used the design and evaluation of videogames?
( ) Smaller ( ) The same ( ) Grater
(3) I would attend other courses that use video games as a learning tool.
( ) Agree ( ) Neutral ( ) Disagree
(4) In general, did you enjoy learning using video games? Explain
( ) Yes ( ) No
(5) Did you find any problem, difficulty, or frustration, in general, when using the design and evaluation of video games as a learning tool? If you had them, list the problems.
( ) Yes ( ) No
(6) If you could improve the experience of using video games in learning HCI, what would you change or add?

References

1. Hewett, T.; Baecker, R.; Card, S.; Carey, T.; Gasen, J.; Mantei, M.; Perlman, G.; Strong, G.; Verplank, W. *ACM SIGCHI Curricula for Human-Computer Interaction*; ACM: New York, NY, USA, 2012.
2. Greenberg, S. Teaching human computer interaction to programmers. *Interactions* **1996**, *3*, 62–76. [CrossRef]
3. Blumenfeld, P.C.; Soloway, E.; Marx, R.W.; Krajcik, J.S.; Guzdial, M.; Palincsar, A. Motivating Project-Based Learning: Sustaining the Doing, Supporting the Learning. *Educ. Psychol.* **1991**, *26*, 369–398.
4. Churchill, E.F.; Bowser, A.; Preece, J. Teaching and learning human-computer interaction: Past, present, and future. *Interactions* **2013**, *20*, 44. [CrossRef]
5. Hartfield, B.; Winograd, T.; Bennett, J. Learning HCI design: Mentoring project groups in a course on human-computer interaction. In Proceedings of the Twenty-Third SIGCSE Technical Symposium on Computer Science Education—SIGCSE '92, Kansas City, MO, USA, 5–6 March 1992; ACM Press: New York, NY, USA, 1992; pp. 246–251.
6. Pastel, R. Integrating science and research in a HCI design course. In Proceedings of the 36th SIGCSE Technical Symposium on Computer Science Education—SIGCSE '05, St. Louis, MO, USA, 23–27 February 2005; ACM Press: New York, NY, USA, 2005; p. 31.

7. Lorés, J.; Granollers, T.; Aguiló, C. An Undergraduate Teaching Experience in HCI at the University of Lleida. Available online: https://pdfs.semanticscholar.org/7789/55598ee05efab7f85aada039363368f31c4b.pdf (accessed on 22 March 2019).
8. Koppelman, H.; van Dijk, B. Creating a realistic context for team projects in HCI. In Proceedings of the 11th Annual SIGCSE Conference on Innovation and Technology in Computer Science Education—ITICSE '06, Bologna, Italy, 26–28 June 2006; ACM Press: New York, NY, USA, 2006; p. 58.
9. Reimer, Y.J.; Douglas, S.A. Teaching HCI Design with the Studio Approach. *Comput. Sci. Educ.* **2003**, *13*, 191–205. [CrossRef]
10. Solano, A. Teaching experience of the human-computer interaction course at the Universidad Autónoma de Occidente of Colombia. In Proceedings of the XVIII International Conference on Human Computer Interaction—Interacción '17, Cancun, Mexico, 25–27 September 2017; ACM Press: New York, NY, USA, 2017; pp. 1–2.
11. Urquiza-Fuentes, J.; Paredes-Velasco, M. Investigating the effect of realistic projects on students' motivation, the case of Human-Computer interaction course. *Comput. Hum. Behav.* **2017**, *72*, 692–700. [CrossRef]
12. Serrano-Cámara, L.M.; Paredes-Velasco, M.; Alcover, C.-M.; Velazquez-Iturbide, J.Á. An evaluation of students' motivation in computer-supported collaborative learning of programming concepts. *Comput. Hum. Behav.* **2014**, *31*, 499–508. [CrossRef]
13. Malone, T.W. *What Makes Things Fun to Learn? Heuristics for Designing Instructional Computer Games*; ACM Press: New York, NY, USA, 1980; pp. 162–169.
14. Entertainment Software Association. The U.S. Video Game Industry's Economic Impact. Available online: http://www.theesa.com/article/u-s-video-game-industrys-economic-impact/ (accessed on 22 March 2019).
15. Dyck, J.; Pinelle, D.; Brown, B.; Gutwin, C. Learning from Games: HCI Design Innovations in Entertainment Software. In Proceedings of the Graphics Interface 2003, Halifax, NS, Canada, 11–13 June 2003; CRC Press: Boca Raton, NM, USA, 2003; pp. 237–246. [CrossRef]
16. Rodríguez-Ortiz, M.A.; Santana Mancilla, P.-C.; Garcia-Ruiz, M.A. El Encanto: Juego Serio para el Aprendizaje de Intervención Comunitaria para Estudiantes de Trabajo Social. In Proceedings of the Memorias del XII Congreso Nacional de Investigación Educativa (COMIE 2013), Guanajuato, Mexico, 18–22 November 2013; Comie: Ciudad de Mexico, Mexico, 2013.
17. Lameras, P.; Arnab, S.; Dunwell, I.; Stewart, C.; Clarke, S.; Petridis, P. Essential features of serious games design in higher education: Linking learning attributes to game mechanics: Essential features of serious games design. *Br. J. Educ. Technol.* **2017**, *48*, 972–994. [CrossRef]
18. Shaffer, D.W.; Squire, K.R.; Halverson, R.; Gee, J.P. Video Games and the Future of Learning. *Phi Delta Kappan* **2005**, *87*, 105–111. [CrossRef]
19. Robertson, J.; Howells, C. Computer game design: Opportunities for successful learning. *Comput. Educ.* **2008**, *50*, 559–578. [CrossRef]
20. Adam, S.; Jakub, S.; Wojciech, A. Computer Game Design Classes: The Students' and Professionals' Perspectives. *Inform. Educ.* **2010**, *9*, 249–260.
21. Dunwell, I.; de Freitas, S.; Petridis, P.; Hendrix, M.; Arnab, S.; Lameras, P.; Stewart, C. *A Game-Based Learning Approach to Road Safety: The Code of Everand*; ACM Press: New York, NY, USA, 2014; pp. 3389–3398.
22. Hertzog, T.; Poussin, J.-C.; Tangara, B.; Kouriba, I.; Jamin, J.-Y. A role playing game to address future water management issues in a large irrigated system: Experience from Mali. *Agric. Water Manag.* **2014**, *137*, 1–14. [CrossRef]
23. Kato, P.M.; Cole, S.W.; Bradlyn, A.S.; Pollock, B.H. A Video Game Improves Behavioral Outcomes in Adolescents and Young Adults with Cancer: A Randomized Trial. *Pediatrics* **2008**, *122*, e305–e317. [CrossRef] [PubMed]
24. Bernhaupt, R.; Isbister, K.; de Freitas, S. Introduction to this Special Issue on HCI and Games. *Hum. Comput. Interact.* **2015**, *30*, 195–201. [CrossRef]
25. Villaroman, N.; Rowe, D.; Swan, B. Teaching natural user interaction using OpenNI and the Microsoft Kinect sensor. In Proceedings of the 2011 Conference on Information Technology Education—SIGITE '11, West Point, NY, USA, 20–22 October 2011; ACM Press: New York, NY, USA, 2011; p. 227.
26. Covaci, A.; Ghinea, G.; Lin, C.-H.; Huang, S.-H.; Shih, J.-L. Multisensory games-based learning—Lessons learnt from olfactory enhancement of a digital board game. *Multimedia Tools Appl.* **2018**, *77*, 21245–21263. [CrossRef]

27. Boyle, E.; Connolly, T.M.; Hainey, T. The role of psychology in understanding the impact of computer games. *Entertain. Comput.* **2011**, *2*, 69–74. [CrossRef]
28. Jiménez, A.S.A.; Ramírez, N.A.B.; Santana-Mancilla, P.C.; Romero, J.C.M.; Acosta-Díaz, R. *Heuristic Evaluation of a Gamified Application for Education in Patients with Diabetes*; ACM Press: New York, NY, USA, 2018; pp. 1–4.
29. Budiu, R. Between-Subjects vs. Within-Subjects Study Design. Available online: https://www.nngroup.com/articles/between-within-subjects/ (accessed on 22 March 2019).
30. Gonzalez, F.; Santana, P.C.; Calderon, P.; Munguía, A.; Arroyo, M. Fallbox: A computer game with natural interaction through head tracking. In Proceedings of the 3rd Mexican Workshop on Human Computer Interaction (MexIHC 2010), San Luis Potosí, Mexico, 8–10 November 2010; Universidad Politécnica de San Luis Potosí: San Luis Potosi, Mexico, 2010.
31. Santana, P.C.; Salazar, J.; Acosta, R. Tangible Multimodal Interaction on a First Person Shooter Game. In Proceedings of the Memorias del I Congreso Internacional de Electrónica, Instrumentación y Computación 2011, Minatitlan, Mexico, 22–24 June 2011.
32. Millen, D.R. Pen-Based User Interfaces. *ATT Tech. J.* **1993**, *72*, 21–27. [CrossRef]
33. Niyazi, K. Mouse Simulation Using Two Coloured Tapes. *Int. J. Inf. Sci. Technol.* **2012**, *2*, 57–63. [CrossRef]
34. Fang, Y.; Wang, K.; Cheng, J.; Lu, H. A Real-Time Hand Gesture Recognition Method. In Proceedings of the 2007 IEEE International Conference on Multimedia and Expo, Beijing, China, 2–5 July 2007; IEEE: Beijing, China, 2007; pp. 995–998.
35. Buxton, B. *Sketching User Experiences: Getting the Design Right and the Right Design*; Morgan Kaufmann: Amsterdam, The Netherlands, 2011; ISBN 978-0-12-374037-3.
36. Buxton, B. *Sketching User Experiences*; Greenberg, S., Carpendale, S., Marquardt, N., Buxton, W., Eds.; Elsevier/Morgan Kaufmann: Amsterdam, The Netherlands; Boston, MA, USA, 2012; ISBN 978-0-12-381959-8.
37. Campbell, R.; Ash, J. Comparing bedside information tools: A user-centered, task-oriented approach. *AMIA Annu. Symp. Proc. AMIA Symp.* **2005**, *2005*, 101–105.
38. Santana-Mancilla, P.C.; Gaytán-Lugo, L.S.; Rodríguez-Ortiz, M.A. Usability Testing of Serious Games: The Experience of the IHCLab. In *Games User Research: A Case Study Approach*; CRC Press: Boca Raton, FL, USA, 2016; pp. 271–283.
39. Desurvire, H.; Caplan, M.; Toth, J.A. Using heuristics to evaluate the playability of games. In Proceedings of the Extended Abstracts of the 2004 Conference on Human Factors and Computing Systems—CHI '04, Vienna, Austria, 24–29 April 2004; ACM Press: New York, NY, USA, 2004; p. 1509.
40. Charlotte, W.; Kalle, J. Heather Desurvire Evaluating Fun and Entertainment: Developing a Conceptual Framework Design of Evaluation Methods. In Proceedings of the INTERACT'07, Rio de Janeiro, Brazil, 10–14 September 2007; Springer: New York, NY, USA, 2007.
41. Santana-Mancilla, P.C.; Montesinos-López, O.A.; Garcia-Ruiz, M.A.; Contreras-Castillo, J.J.; Gaytan-Lugo, L.S. Validation of an instrument for measuring the technology acceptance of a virtual learning environment. *Acta Univ.* **2019**, *29*, e1796.
42. Pedro, C.; Santana-Mancilla, M.A.; García-Ruiz, L.S.; Gaytán-Lugo, M.A.; Rodríguez-Ortiz, S.B. Fajardo-Flores Uso de Juegos Serios para la Enseñanza-Aprendizaje de Competencias en Nivel Superior. In *Interacción Humano-Computadora y Aplicaciones en México*; Luis, A., Marcela, C., Rodríguez, D., Eds.; Academia Mexicana de Computación: Ciudad de Mexico, Mexico, 2018; pp. 193–212.

© 2019 by the authors. Licensee MDPI, Basel, Switzerland. This article is an open access article distributed under the terms and conditions of the Creative Commons Attribution (CC BY) license (http://creativecommons.org/licenses/by/4.0/).

Article

FLIGBY—A Serious Game Tool to Enhance Motivation and Competencies in Entrepreneurship

Zoltan Buzady [1] and Fernando Almeida [2,*]

[1] Corvinus Business School, Corvinus University of Budapest, 1093 Budapest, Hungary
[2] Faculty of Engineering, University of Porto & INESC TEC, 4200-465 Porto, Portugal
* Correspondence: almd@fe.up.pt

Received: 27 June 2019; Accepted: 17 July 2019; Published: 19 July 2019

Abstract: Entrepreneurship is currently one of the most fundamental economic activities in the 21st century. Entrepreneurship encourages young generations to generate their self-employment and develop key soft-skills that will be useful throughout their professional career. This study aims to present and explore a case study of a higher education institution that adopts FLIGBY as a serious game, which allows students to develop entrepreneurship skills in an immersive way and based on real challenges that can be found in business environments. The findings indicate that FLIGBY offers relevant potentials and new possibilities in the development of management, leadership, and entrepreneurship skills. Furthermore, the game allows the inclusion of summative and formative assessment elements, which are essential in the process of monitoring and analyzing the student's performance.

Keywords: entrepreneurship education; FLIGBY; Flow; positive psychology; serious games; higher education

---

1. Introduction

In the past, it was believed that entrepreneurship could not be taught. In this sense, entrepreneurs were individuals who had an innate capacity, having been born with special characteristics that favored business success. However, the results of several studies over the last two decades have shown that although personal characteristics may facilitate the management of a new business, the entrepreneurial process can be learned and trained [1,2]. According to [3], business success will depend on several internal and external factors to the business, on the entrepreneur's personal characteristics, and on his/her ability to manage daily challenges.

One of the main concerns of entrepreneurial education is to build active entrepreneurs, going beyond the theoretical knowledge of the theme [4,5]. However, Lautenschläger and Haase [6] emphasize that there are aspects of entrepreneurship that may be easy to teach and others not. Skills and competencies such as creativity, innovation, pro-activity, decision-making, and risk propensity are difficult aspects to instill in individuals without this intrinsic propensity [6]. Several authors such as [7,8] advocate a more practice-oriented pedagogical approach as more appropriate for the teaching of entrepreneurship. Therefore, the traditional lecture can be used to review theoretical and cultural aspects of entrepreneurship, directing the other aspects of entrepreneurial action towards richer, more dynamic, and immersive methods and pedagogic resources.

The teaching of entrepreneurship should not exclusively teach how to run a business. In [9], it is mentioned that this type of teaching should encourage creative thinking and promote a strong sense of self-esteem and autonomy. The knowledge that should result from the teaching of entrepreneurship should promote communication, creativity, critical thinking, leadership, negotiation, problem-solving, networking, and time management [10,11]. According to [12], the teaching of entrepreneurship helps to raise students' awareness of the entrepreneurial career, transferring general knowledge about business management and increasing the level of knowledge and empowerment of students to create their own professional career.

In entrepreneurial education programs, traditional methods are based on theoretical lectures and group work. This approach allows students to understand rigorously and deeply the characteristics of entrepreneurial activity [13]. However, this approach does not allow students to understand the consequences of the actions taken before the commitments and resources to launch a business [14]. Consequently, other teaching methods should complement this theoretical training and encourage the active participation of students, the enhancement of their social and problem-solving skills. Belitski et al. [15] advocate formal education should be complemented by informal teaching methods supported in group activities, simulations, and serious games. In fact, a serious game is defined by [16] as games that have an educational objective but also include three fundamental components: (i) experience, (ii) entertainment, and (iii) multimedia. In this sense, this study intends to explore, through a case study approach, how the use of FLIGBY as a serious game can be used to enhance motivation and competencies in entrepreneurship. This study uses a quantitative approach to assess the success of FLIGBY adoption in an entrepreneurship discipline attended by multidisciplinary students from the management and computer science course at a polytechnic higher education institution in Portugal.

The work is organized as follows: Initially, a literature review on the field of gamification and serious games is performed. After that, the FLIGBY game and the concept of Flow theory is presented. Thereafter, the methodology and adopted methods are presented. Consequently, the main findings are analyzed and discussed. Finally, the conclusions of this work are drawn and some suggestions for future work are given.

2. Background

Games are typically designed with the primary objective of providing a playful experience and entertaining players. Studies conducted by [17,18] show good results in the application of games to increase the motivation of players when used to make products and services more attractive and engaging consumers. The use of game mechanics, dynamics, and structure in non-game contexts is called gamification [19].

Gamification has been incorporated into several businesses in the promotion of marketing strategies and customer engagement. Deterding et al. [19] address the use of reward systems based on users' reputation ranking. This reward-based approach was then used by [20] to define a model that awards points according to customer participation (e.g., register in products, submit comments and reviews, watch videos, Facebook "likes", share on Twitter, etc.). Edwards et al. [21] suggest the use of gamification techniques incorporated in smartphone applications in the promotion of health behaviors. Due to its popularity, gamification extrapolated the business segment and began to be applied in other contexts such as education.

The purpose of gamification in education is to make learning easier, faster, and more dynamic for the student. According to [22], theoretical content alone does not capture the attention and interest of students in the subjects addressed, hence the retention of this content is low. In this sense, gamification has the fundamental idea of awakening the greater involvement of students and, consequently, improving the quality of teaching. Sailer et al. [23] state the potential of gamification in education is immense, functioning as an element that arouses interest, increases participation, and develops creativity and autonomy. To make use of gamification, it is not necessary to use a game despite the fact that it is one of the possibilities. Gamification in education comprises the idea of adding elements, mechanics, and logic of games to engage people in the learning process [24].

Simultaneously with the growing interest in gamification, the concept of serious games emerged. According to [25], serious games are designed for a specific purpose related to learning, not just for fun. They have the characteristic elements of a playful game, but their goal is to counsel something predetermined and related to the learning process. Karagiorgas and Niemann [26] state that serious games require players to meet goals in a virtual world, whose student performance has no consequence in the real world, while gamification in education is created to increase student engagement in real tasks.

The concept of serious games is an emerging paradigm in Technology Enhanced Learning (TEL). The characteristics of serious games allow their adoption in the development of entrepreneurial skills and for the teaching of entrepreneurship [27,28]. According to [29], serious games are effective because the acquisition of knowledge is done using a contextual learning approach. Hauge et al. [30] complement this vision by emphasizing that good entrepreneurship serious games should involve students in challenging environments that simultaneously motivate the acquisition of skills relevant to entrepreneurial activity and promote the development of soft skills.

The adoption of serious games in entrepreneurship classes is quite residual. One of the most relevant works in the area was carried out by [31], in which a comparative analysis of eight entrepreneurship serious games (i.e., GoVenture: Entrepreneur, GoVenture: World, Interpretive solutions: Entrepreneur, Entrepreneurship Simulation: The Startup Game, SimVenture, HipsterCEO, Innovative Dutch, and Venture Strategy) was performed. The findings of this study are aligned with the results also obtained by Bellotti et al. [32] and highlight the low degree of maturity of these projects, and the low level of fidelity and engagement. Another game that stands out in the European paradigm is the ENTREXplorer. The main objective of the game is to offer a technological solution, in which students can learn the fundamental concepts of entrepreneurship, the creation of a new business, marketing, and strategic positioning [33]. This game also allows students to build a business plan online. Finally, the work developed by Almeida and Simões [28] identified that entrepreneurship serious games typically fail to support different learning styles and do not include an internal process of assessing learning goals. Therefore, it is important to explore the potentialities of FLIGBY and assess whether the use of FLIGBY increased the students' motivation to attend the entrepreneurship discipline. The following research question (RQ) was established: *RQ1: Is the motivation to attend entrepreneurship classes increased?*

Following the recommendations of [7,8], it is essential to offer students a practical environment for the teaching of entrepreneurship that complements the theoretical exposition of the theme of entrepreneurship. The literature highlights the use of group work in which students can simulate the process of developing a business plan that includes aspects related to the generation of ideas, strategic positioning, feasibility analysis, among others [34,35]. In the process of building a business plan, it is essential that students work in groups as a way to promote communication, the active learning process and the development of critical thinking. Rosen [36] emphasizes the fundamental role of teamwork for entrepreneurs in the maturing of ideas, mentioning that the courses of the main university institutions include team-based project work. In this sense, it is also important to explore whether the use of FLIGBY contributes to increasing students' perception of the importance of group work. The following research question was included: *RQ2: Is the perception of the importance of group work increased?*

The process of developing serious games is based on a diverse set of educational theories, such as cognitive, constructivist, constructionist, and experimental theories of learning. These theories are based on the pillar of contextual learning because the information used in this context tries to simulate reality, and it is supposed to transfer the player to that reality. Based on the culture and identity of each player, serious games can be a way to mediate learning through discussion and reflection [37]. The application of serious games to teaching shows that the motivation before the game, their involvement, and satisfaction while playing depends on the level of their effort and the previous preparation performed by the teacher [38]. This recommendation is important in the definition of mechanisms to introduce students to the fundamental theoretical concepts associated with FLIGBY and to monitor the difficulties experienced by students throughout the game.

Serious games offer the educational community more realistic and interactive educational processes. The educational approach of the entrepreneur based on serious games makes possible to unite playful aspects and specific content, motivating the learning process and contributing to make students feel better prepared for the challenges of creating a new business. The study conducted by [39] shows that entrepreneurship education has contributed to increasing the competencies and intentions of students toward self-employment. Other studies have emerged to explore the type of skills that entrepreneurship education promotes. In [40], the types of competences developed are clustered into two major groups: Academic competencies and personal effectiveness competencies. The first cluster includes competencies regarding reading, writing, science, technology, or critical and analytical thinking; the second cluster includes interpersonal skills, initiative, ambition, integrity, or taking risks. This mapping is not unique, emerging studies that include other groups. In [41], it is suggested the adoption of three clusters: Entrepreneurial attitude (e.g., sense of initiative, self-efficacy, structural behavior), entrepreneurial skills (e.g., creativity, networking, adaptability), and knowledge of entrepreneurship (e.g., understanding of entrepreneurship, role of entrepreneurs, determinants of successful entrepreneurship). In this sense, it is important to explore the role that FLIGBY can play in the development of an entrepreneur's key competencies. The following research question was included: *RQ3: What are the skills in which students performed better or worse?*

The assessment of learning through the use of serious games is another key element. Yüksel and Gündüz [42] emphasize the complementary role of formative and summative assessment. Summative assessment refers to the students' performance at the end of a course or subject in order to assess the final outcome of the learning process, assigning grades and certifications, while formative assessment seeks to monitor the student's learning process throughout the use of a method or tool in teaching [43]. Price [44] states formative assessment is continuous and systematic, allowing the teacher and students to obtain information about the development of learning with a view to adjusting processes and strategies. Serious games should implement final and intermediate assessment mechanisms to assess students' performance. Accordingly, Bellotti et al. [45] advocate that the evaluation of students' performance should consider the different learning styles and provides appropriate feedback throughout the game. Daoudi et al. [46] summarize the evaluation techniques that can be used for learners' assessment, namely the use of interviews, questionnaires, scoring, and multi-agent systems. Within these, the adoption of surveys and questionnaires stands out due to its simplicity and independence from the type of game adopted. However, it is clear that these techniques can be effectively complemented and used in conjunction with mechanisms integrated into the game to measure the performance of the game in real-time. Therefore, it is important that a pedagogical strategy for the use of serious games in the development of entrepreneurial skills includes summative and formative components. Consequently, the following research question was included: *RQ4: What are the summative and formative elements used in the assessment of students' performance?*

3. The FLIGBY Serious Game

3.1. The Flow Theory

The flow theory, proposed by Mihaly Csikszentmihalyi from his initial research on states of happiness and creativity, advocates that happiness occurs when individuals experience optimal mental states called "Flow". This state of immersion in which people entered when devoting themselves to a particular task can be achieved from the existence of three conditions [47]:

1. The activities in which the individual is involved have clear objectives and provide an explicit way of reporting on progress in them;
2. Feedback needs to be immediate, whether positive or negative;
3. The skills needed to perform an activity and the degree of difficulty of the proposed exercise need to be in perfect balance.

The various emotional states proposed in [47] in which an individual can be while performing an activity are summarized in Figure 1. Depending on the skill level of the individual and the difficulty of the challenges, the individual will be in a different mental state. The "apathy" is considered the worst state, where there is no challenge to meet and no skill to test. The opposite is exactly the state of flow, where the individual puts all his or her skills into the realization of a challenge that he or she considers both challenging and capable of accomplishing. Most of the time, it is expected that an individual can be at intermediate points. For example, if an individual is in a state of control, then it is important to increase the challenge; if he/she is in a state of anxiety or excitement, then it is important to increase his/her abilities to perform a proposed task.

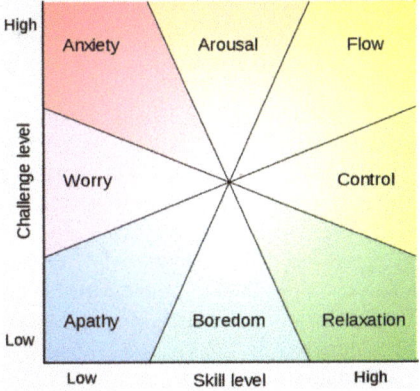

Figure 1. Emotional states in the Flow theory [47].

In the research conducted by [47], which involved a very diverse group of professionals (e.g., monks, executives, doctors), some common elements were found that pointed to the state of flow, respectively: (i) complete involvement in the activities developed, (ii) high focus and concentration, (iii) internal clarity, (iv) calm mind and no worries, (v) total focus on the present moment with loss of the notion of time duration, and (vi) intrinsic motivation.

3.2. Architecture and Structure of FLIGBY

FLIGBY is a serious game in which the player is recently named General Manager (GM/CEO) of a family business ("Turul Winery") in California in the USA. The player faces the challenging task of having to achieve a state of harmony and cooperation in a team significantly weakened by internal conflicts due to the dysfunctional leadership style of the previous GM.

A key task is to create an environment that promotes teamwork and improves Flow. Consequently, one of the main goals of the game is to bring as many colleagues as possible, even for a short time, to a state of Flow. The dilemma is when to support a colleague and when to protect the interest of the team, its stakeholders, and prudent environmental management. At the same time, decisions about strategic issues of the company's future must be made according to the expectations of the owner of the winery.

The player performs approximately 150 decisions during the game. In most decisions, for example on how to conduct a strategy meeting with a team, the player must choose a response from 2 to 5 options presented. The selected answers will put each player in a proper and individual path of the story. There are many possible paths that result in different results at the end of the game. The experience of using the game is therefore personalized by the player according to the decisions made throughout the game.

The player will necessarily have to balance several goals to employees reach the Flow state. This can be done in two ways: By making management decisions that affect colleagues so that they can enter a state of Flow or by calibrating their decisions so as to promote a corporate atmosphere conducive to Flow overall. It is also the winery's goal to follow business practices that are friendly to the wider social and ecological environment, without compromising its profit potential.

The overall performance of the player as GM of Turul Winery will be indicated whether or not he wins the maximum prize of the game, the so-called "Spirit of Wine Trophy". The average playing time is 7–9 h. The software structure consists of the following three main parts:

1. The game action—this is the game itself, presented in an interactive movie format, in which the player is advancing through the story according to his/her choices and decisions;
2. The database—stores all data (login, decisions, choices, actions) generated during the game;
3. The Master Analytics Profiler (MAP)—a final report for each individual player is generated, in which the player's performance is measured considering 29 key management and leadership skills. Each of these competencies is further detailed in the methodology section.

In the implementation of FLIGBY were considered the four stages of the approaches for serious game design (i.e., analysis, design, development, and evaluation) synthesized by Ávila-Pesántez et al. [48]. FLIGBY offers the possibility of being used in different domains. This flexibility is achieved through the presentation of a wide variety of scenarios and situations, which allows the game to be explored in an academic and business environment with individuals in the field of management, leadership, or entrepreneurship. According to [49], this high flexibility is fundamental for the commercial success of a serious game in different industries. Additionally, FLIGBY was built with the objective of building learning in a constructivist and experiential way. In line with [50,51], this is a good practice that should be followed by serious games to remove any useless redundancy and engage the player throughout the various challenges that are posed by the game.

The game is available through a web-browser interface. The dynamics of the game occur in the central window where the various characters interact with each other. In the lower left corner, the game presents the consultant persona (Mr. FLIGBY), who is a key character that gives feedback to the player on his/her actions (see Figure 2). On the right side, the game offers several buttons to access the features of the game and also gives an indication regarding the time spent in each scenario. Marer et al. [52] describe many details about FLIGBY's conceptual, academic, and pedagogical aspects.

Figure 2. FLIGBY interface (authors' own illustration).

4. Materials and Methods

4.1. Structure of the Study

Mixed methods were employed in the development of this study. This approach allows us to simultaneously use quantitative and qualitative methods in the exploration and analysis of data. According to [53], the adoption of mixed methods enriches the findings and increases the depth and breadth of the study. The literature offers several ways to structure a mixed methods study, and the multilevel design was applied in this study. This approach assumes the existence of multiple layers in the analysis of a problem, in which quantitative and/or qualitative methods can be used throughout the process. In [54] it is stated that this approach allows us to explore the cross-level direct effects and provide more accurate and robust results. Figure 3 represents schematically the adopted multilevel design in which it becomes possible to perceive the objective of each phase and the adopted method. The study was structured in three phases, with qualitative and quantitative techniques being adopted in the collection and analysis of data in each phase. The first phase followed from 15 October 2018 to 1 November 2018; the second phase from 1 November 2018 to 15 December 2018; and the third phase from 7 January 2019 to 25 January 2019.

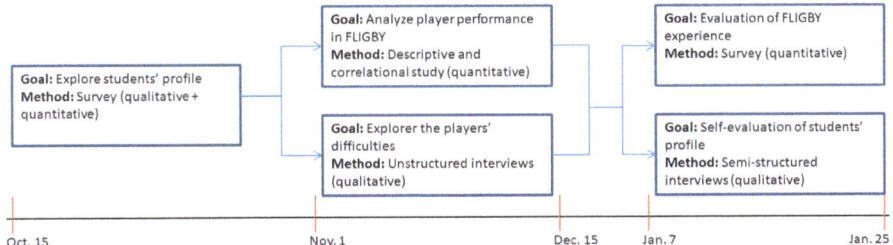

Figure 3. Phases of the adopted methodology (author's own illustration).

The study began on 15 October 2018, in which a survey was distributed among students that aimed to characterize the students' profile considering their possible previous experience as entrepreneurs, the experience of working in groups, their individual attitudes, and the self-evaluation of their skills. This survey was available to students during the 2-week period. Among the various methodological options in the field of social sciences and humanities research, the survey is assumed to be one of the most frequently used. According to [55,56], the use of surveys allows direct knowledge of reality and rapid quantification of the findings. However, it is also an approach with low perception and comprehension of evolutionary phenomena. For these reasons, we adopted a research methodology consisting of three phases (Figure 3) in which qualitative and quantitative methodologies were used along the research process. Table 1 presents a brief descriptive analysis of the survey. A total of 51 students responded to this survey. The results of this survey show that a significant majority of students do not have experience as entrepreneurs and consider that they enjoy working in groups. A considerable majority of learners accepted failures in their learning process and considered that these situations offered opportunities for improvement. In relation to quality, timing, and perfectionism there was a greater dispersal of responses, and there was no clear conviction from students as to which of these elements may be most important.

This survey also contained some qualitative interpretation questions that aimed to explore the process of working in a group and self-evaluate students' skills, respectively:

- Students mentioned that in group working is easier to divide tasks and increase productivity. Additionally, group working contributes to a large amount of knowledge sharing, increases levels of creativity and enables new approaches to be adopted in solving challenges. On the other hand, the main challenges are communication, the difficulties in generating consensus and inadequate/unequal participation of the members of the group;
- Students believed that their core competencies include the ability to organize, execute and engage in projects. On the opposite side, communication, time management skills and emotional intelligence were pointed out as the skills in which they expected to have the greatest difficulties.

In phase II which took place between 1 November 2018 and 15 December 2018 the FLIGBY was made available to students via the cloud. An individual access account was created for each student. The student's performance during the game was recorded. During these 6 weeks, a weekly session of 30 min was organized to discuss the main difficulties experienced by students in the game until that week. An informal approach was chosen through unstructured interviews with the students of the class. Only two types of problems were found: (i) the application did not work on IoS devices due to technical problems in running a Flash application; and (ii) due to network breaks, some students had to play the same scenario several times. These problems were overcome with the support of the teacher. These sessions also proved to be essential to increase the students' level of motivation for the game and to ensure that all students received the access code for the game.

Table 1. Sample characteristics of the survey (phase I).

Variable	Absolute Frequency	Relative Frequency
Have you had experience as a manager?		
Yes	5	0.098
No	46	0.902
Have you already founded your own company?		
Yes	1	0.020
No	50	0.980
Do you like to work in a group?		
Yes	49	0.961
No	2	0.039
Individual Attitudes	**Mean**	**Std. Dev.**
I can accept failures as part of a learning process	4.255	0.595
Failures allow opportunities for reflection and consideration	4.353	0.595
I quickly overcome setbacks	3.647	0.716
I feel comfortable to talk to people that are different from me	4.137	0.825
I frequently come in contact with people that are different from me	3.725	0.981
When facing difficult tasks, I am certain I will accomplish them	3.882	0.653
I would prefer to hand in a product on time rather than making it perfect	3.255	0.997
In general, quality and perfection are more important than effectiveness	3.137	1.000
I will probably start my own business one day	3.686	0.761

In phase III, the authors tried to evaluate the experience of using FLIGBY and the performance obtained by the students. This process was performed using a survey in which quantitative data on the experience of the game, its benefits, and limitations were obtained. Table 2 shows the data obtained. In the "evaluation of experience" dimension, the students gave a high score for the experience offered by FLIGBY, having also mentioned that it was an experience "worth the time". They also considered that the game is not too long or complex, despite a high dispersion of responses. In the "evaluation of game" dimension, the students pointed to interactivity as the main asset of the game. In the "benefits of the game" dimension, the students considered that FLIGBY allowed them essentially to increase the level of knowledge in the management field and to have a greater knowledge of their skills that will be essential to them along their career at university. On the other side, in the "limitations of the game" dimension, the students considered that FLIGBY needs previous training before starting to play. In this sense, the pedagogical session held by the teacher one week before the game was played by the students proved to be absolutely essential. This session lasted 90 min and two topics were covered: (i) presentation of the flow theory and how this theoretical concept is essential in the management of an organization and team, and (ii) explanation of the interface of the game.

In addition, a semi-structured study was performed in which students were asked to perform a self-assessment of their competences against the performance indicators provided by the game. This semi-structured interview was conducted in a group in which all students were present and three questions were asked: (i) the main difficulties experienced by students in using FLIGBY, (ii) the indicators in which they performed best and worst were in line with their expectations, and (iii) what is the influence of their performance in FLIGBY on their academic and professional career.

Table 2. Sample characteristics of the survey (phase III).

Variable	Mean	Std. Dev.
Evaluation of experience		
Please rate your overall FLIGBY experience	4.105	0.649
Please rate: The FLIGBY experience was "worth the time."	4.053	1.064
What is your opinion: The FLIGBY game is "too long"	2.579	1.130
What is your opinion: The FLIGBY game is "too complex"	2.789	1.143
After playing FLIGBY. Would you like to open your own business?	3.289	0.956
After playing FLIGBY. Do you feel you have more skills to open your own business?	3.684	0.873
After playing FLIGBY. Did you get more competencies to work in a group?	3.868	0.777
Evaluation of game		
Realism	3.868	0.811
Engagement	3.763	0.820
Immediate feedback	3.816	0.926
Interactivity	3.974	0.716
Personalization	3.553	0.795
Benefits of the game		
Improves knowledge in the management field	4.105	0.509
Improves knowledge in the leadership field	4.079	0.587
Improves knowledge in the entrepreneurship field	4.053	0.613
Helps me to be more aware of my skills and actions at work/university	4.105	0.649
Helps me to try out new approaches	4.079	0.587
Helps me to have new attitudes to people	4.053	0.655
Improves my self-esteem	3.474	0.797
Improves collaborative learning	3.816	0.652
Enhance my motivation	3.684	0.842
Helps to know more about myself	3.763	0.820
Applicable to the real world	3.842	0.789
Improves to establish new social connections	3.553	0.828
Limitations of the game		
Requires high training before starting to play	2.737	1.057
Decreases students' attention to the classes	2.184	1.159
Creates isolation feelings on the students	2.237	0.971
Evaluation not related to the course assessment	2.421	1.106
Decreases the time available to focus on the classes	2.342	1.146

A thematic analysis was employed to analyze the findings of the semi-structured interview (Table 3). The thematic analysis is one of the most commonly used approaches in qualitative research methods and, according to Braun and Clarke [57], allows identifying and exploring patterns within the data. Furthermore, Creswell and Poth [58] point out that this approach allows exploring the data from two perspectives: From a data-driven perspective and from the research question perspective. This complementarity turns possible to assess the consistency of the data with the research questions and to explore the level of detail obtained in each of them.

Table 3. Identification of final themes in the thematic analysis process.

Dimension	Sub-Dimension	Final Themes
Experienced difficulties	Technical	Compatibility issues with all browsers Low bandwidth of the Internet connection
	Scientific	Achieve the Flow state at the end of the game Dealing with conflicts and heterogeneity
	Pedagogical	Dedicated time to the entrepreneurship discipline
Indicators	Best	Emotional intelligence Social dynamics Information gathering Better performance of computer science students in five MAP dimensions
	Worst	Time management Decision-making on pressure Delegation Prioritization
Influence	Academic	Development of management, leadership and entrepreneurship skills Reducing knowledge asymmetries More receptive to working in groups Self-assessment tool
	Professional	Increase motivation to work with multidisciplinary teams
		Development of soft-skills that are essential in the labor market

4.2. Research Dimensions

Player performance in FLIGBY is measured using 29 indicators. These indicators correspond to the skills identified by [47] from interviews conducted with CEOs and organizational leaders, which allowed the leader persons to personally reach the state of flow, in which there is total absorption and joy in the present moment, and also to create what Csikszentmihalyi calls "Flow-promoting organizational work environments and leadership culture". Therefore, in the business context, in order to obtain the best from each employee and to obtain business success, it becomes necessary to have harmony and positive results in each of the 29 indicators identified in Table 4. FLIGBY creates an individual final report for each player in which its performance is measured considering those 29 indicators. This report is made available to the teacher and to each student.

Table 4. MAP dimensions

Dimension	Description
Active listening	How to respond to another person who improves mutual understanding. It involves understanding the content of a message, as well as the sender's intention and the circumstance under which the message is given.
Analytical skills	Ability to visualize, articulate, and solve complex problems and concepts and make decisions that are sensible on the basis of available information.
Assertiveness	Ability to express their emotions and needs without violating the rights of others and without being aggressive.
Balance	Ability to maintain the same importance between things. A balance between challenges and skills is necessary for Flow.
Business-oriented thinking	Ability to manage situations and solve problems in order to create added value for the company.
Communication	Set of skills that enable a person to convey information in order to be received and understood.
Conflicts management	The practice of identifying and dealing with conflicts in a sensible, fair, and efficient manner.

Table 4. Cont.

Dimension	Description
Creation of trust	Ability to create trust and a positive state of mind in which individuals feel the desire to participate.
Decision-making on pressure	Readiness to form, facilitate, and monitor teamwork and teams.
Delegation	Ability to delegate function or authority on another person to act on behalf of the manager.
Diplomacy	Ability to take into account the varying interests and values of other parties involved in the negotiation.
Emotional intelligence	Ability to recognize and evaluate your own and others' feelings.
Empowerment	Competencies to share information, rewards, and power with employees so that they can take initiatives and make decisions.
Entrepreneurship	Ability and willingness to undertake the design, organization, and management of a productive enterprise with all inherent risks, while seeking profit as a reward.
Execution	The act of successfully executing and completing management tasks.
Feedback	Give response, whether positive or negative, to a particular request, action, or event.
Future orientation	Willingness to think long-term and about the future consequences of taken actions.
Information gathering	Willingness to collect appropriate information to carry out the next step on the basis of this information.
Intuitive thinking	Blurred, non-linear way of thinking that does not use rational processes such as facts and data.
Involvement	Readiness to participate in the activities of a group or team.
Motivational skills	The type of skills that enable a person to become motivated and work towards achieving goals.
Organizing	Ability to organize itself around a concept or model that enables the implementation of actions taken in a sustained manner.
Personal strengths	Recognize and apply personal strengths is the ability to discover and use well the personal and other people's strengths that are not immediately obvious.
Prioritization	Ability to organize a set of items and set priorities among them.
Social dynamics	Awareness of the complexity of many situations and the social dynamics that govern them.
Stakeholder management	Ability to manage the business process, usually involving a trade-off, in order to have a positive impact on the organizations' stakeholders.
Strategic thinking	Ability to think that allows the discovery of alternatives with considerable effectiveness in achieving an objective or solving a problem instead of resorting to obvious choices.
Teamwork management	Ability to participate, facilitate, and monitor teamwork.
Time management	Process of planning and executing, being aware of the time allocated, its priority and of the eventual existence of competing activities.

5. Results and Discussion

5.1. RQ1—Is the Motivation to Attend Entrepreneurship Classes Increased?

The discipline of entrepreneurship is attended by students from management and computer science courses. This heterogeneous profile allows the creation of a multidisciplinary learning environment as suggested by the Organization for Economic Cooperation and Development (OECD) when analyzing the state of entrepreneurship education in higher education institutions [59]. However, the process of integrating multidisciplinary students also presents some challenges, namely in terms of communication within the teams, in the division of tasks, and also in the motivation to work together. According to the data available in Table 2, the use of FLIGBY before the formation of these multidisciplinary teams allowed students to know their skills and develop management skills. Both dimensions present a mean equal to 4.105. The semi-structured interview held after the conclusion of the game by students allowed us to identify that the development of management skills was more valued by the students of the computer science course. For the students of the computer science course, the entrepreneurship course is the first curricular unit in which they address basic knowledge of management of an organization, while the management students already attended two disciplines (i.e., Management and Organization of Companies, and Marketing) that introduce fundamental management competencies. Therefore, FLIGBY emerges as a valuable tool in reducing asymmetries in the preparation of students in the creation of an entrepreneurship project.

Looking at Table 2, the authors found that the increase in motivation was positively evaluated by the students, despite being one of the least relevant dimensions within the tested benefits. Nevertheless, motivation cannot be seen in isolation. Frese and Gielnik [60] advocate that there is a slight but important difference between people who seek to perform activities to meet their strengths, which is natural because they feel competent, and the pleasure of facing challenges and testing new skills and which are typical for entrepreneurs. FLIGBY seeks to exactly pose new challenges to students throughout various scenarios. In third and fourth place as the main benefits offered by FLIGBY (see Table 2) stand the development of leadership competencies and the testing of new approaches. According to [61], entrepreneurship and leadership present conceptual similarity with a considerable area of overlap but are still different. While leadership is more directly associated with the conceptual components related to people, entrepreneurship tends to be more linked to the concepts of seeking independence through the exploration of market opportunities. Kouzes and Posner [61] state entrepreneurship is not a necessary part of successful leadership, but leadership is an element of entrepreneurial success. Despite the importance of leadership, there is no discipline in the management course and computer science course that addresses this theme. Students in general, and in particular those from the management course and with experience in the labor market, expressed as a positive point that FLIGBY allowed them to explore how different followed options have an impact on the cohesion of teams and in the individual performance of employees. Two of these students expressed that the knowledge acquired has a direct impact on their business activities as they hold middle management positions, one in a financial company and another in the logistics field.

5.2. RQ2—Is the Perception of the Importance of Group Work Increased?

The development of entrepreneurial projects requires the research, identification, and development of creative and innovative ideas. Group work emerges as a fundamental competence in the various phases of an entrepreneurship project, from the generation of the business idea, conceptualization, to the entry into the market. The first survey made available to students before they started the game allowed us to identify the students' perception of group work. The results confirmed the students' appreciation of this topic since more than 95% of the students said they liked group work and were able to identify some of the main advantages of collaborative work, such as the appreciation of each individual, the development of skills, or the exchange of experiences. The data in Table 1 also allowed us to identify that the majority of students (mean equal to 4137) stated that they felt comfortable in communicating with people different from them, although the number of experiences they had in this situation was relatively lower (mean equal to 3725). In fact, the discipline of entrepreneurship proved to be fundamental in increasing this indicator, since teamwork is absolutely essential and the 2018/19 school year was attended by four Erasmus students from partner institutions in Lithuania and Poland.

The results of the survey available in Table 2 and the semi-structured interview conducted after the conclusion of the game were inconclusive regarding the direct benefits offered by FLIGBY for the increase of group work skills. However, the results of the survey in Table 2, namely, "helps me to have new attitudes to people" with an average of 4.053, show that students after playing FLIGBY feel more receptive to working in groups. Furthermore, students indicated that in their role as CEO of Turul Winery in the FLIGBY game they needed to identify the key skills of each employee and get them to work as a team. FLIGBY's main challenge was to get all Turul Winery employees to achieve Flow by the end of the game. The way each student managed the conflicts within the team allowed them to be better prepared for the real challenges they will encounter in multidisciplinary group work in the context of the entrepreneurship course.

5.3. RQ3—What Are the Skills in Which Students Performed Better or Worse?

Table 4 allows us to explore the students' performance in FLIGBY considering the established 29 MAP dimensions. The students' performance in each dimension was analyzed considering the course in which they were enrolled, and it was possible to reach the following conclusions: (i) the three dimensions in which the students of the business course performed best were information gathering, involvement, and emotional intelligence; (ii) in the computer science course, a similar behavior was observed, plus the importance of social dynamics; (iii) the variables in which the students performed worst were time management, decision-making on pressure, delegation, and prioritization. The dimensions in which the students had the worst performance are globally transversal to both courses.

Table 5 also seeks to assess whether the behavior of students in any of the considered courses is different for each of the 29 MAP dimensions. For this purpose, the p-value was calculated, which represents the probability of having observed a distinct significant value under the null hypothesis. A cut-off value of 0.05 was defined to reject the null hypothesis. The results indicate the existence of different significant performance for the students of the computer science course in the following dimensions: (i) decision-making on pressure; (ii) feedback; (iii) future orientation; (iv) social dynamics; and (v) time management. In these five situations, it was found that the performance of students from the computer science course was significantly higher.

Table 5. Students' performance in FLIGBY considering MAP dimensions.

Variable	Business Course (n = 18)			Computer Science Course (n = 31)			p-Value
	Mean	Median	Std. Dev.	Mean	Median	Std. Dev.	
Active listening	68	71	11	67	71	10	0.6121
Analytical skills	63	65	13	65	67	10	0.5304
Assertiveness	58	58	8	59	59	10	0.7283
Balance	67	66	9.7	67	70	9.3	0.9236
Business-oriented thinking	59	59	7.1	62	64	8.6	0.1455
Communication	63	63	8.9	65	64	11	0.3803
Conflicts management	59	58	6.7	61	63	8.3	0.1963
Creation of trust	68	68	6.5	69	70	7	0.3603
Decision-making on pressure	52	49	9.2	57	56	11	0.0299
Delegation	53	50	11	53	50	18	0.8790
Diplomacy	64	65	12	68	65	10	0.2033
Emotional intelligence	70	73	7.5	73	74	8.7	0.1153
Empowerment	65	67	15	66	67	12	0.7049
Entrepreneurship	64	64	9.7	68	67	12	0.1398
Execution	61	65	13	67	65	10	0.0813
Feedback	66	65	8.2	72	71	16	0.0073
Future orientation	65	63	8.8	70	70	11	0.0169
Information gathering	71	71	11	72	71	9.7	0.6650
Intuitive thinking	61	59	7.6	62	60	8.3	0.6257
Involvement	71	70	11	71	70	9.4	0.8947
Motivational skills	68	70	8.7	69	68	11	0.6317
Organizing	64	65	9.6	64	65	12	0.9418
Personal strengths	65	64	8.9	67	68	10	0.3667
Prioritization	54	53	8.2	56	53	10	0.3126
Social dynamics	66	68	6.6	73	74	9.4	0.0003
Stakeholder management	65	67	12	70	70	15	0.1065
Strategic thinking	62	62	6.9	64	65	10	0.2616
Teamwork management	62	64	7.7	65	65	11	0.1359
Time management	47	50	9.5	52	50	16	0.0256

Shaded cells have a p-value less than 0.05.

The semi-structured interview conducted after the conclusion of the game made possible to explore the reasons that justify the obtained results. Difficulties in time management was a top difficulty pointed out transversally by all students, and which occurred right in the first scenarios of the game in which the player in the role of CEO had to schedule meetings with all employees of the company. In fact, most of the students showed interest in knowing in depth each employee but could not manage the time allocated to each meeting, having typically exceeded this time. This difficulty experienced in the game is also experienced by students in the discipline of entrepreneurship since this discipline works in parallel with other disciplines with a high taught time and workload on projects. In fact, time management is considered a very important determinant in the individual performance of students in higher education, but it is also important for entrepreneurs in order to complete the pre-startup phase successfully [62,63]. On the other hand, the students' performance in the emotional intelligence component was a surprise. Most students were able to identify, understand, recognize, and manage their emotions and the emotions of other characters in the game in a positive way. This positive performance also represented a broad growth of students in this field since most of them did not even know the concept of emotional intelligence. In [64], it is highlighted the importance of developing emotional intelligence in education and FLIGBY is a non-immersive solution that can promote the development of these skills.

Finally, the authors also explored the reasons why computer science students got better results in five MAP dimensions. At this level, three reasons emerged as relevant to explain this performance: (i) project management practices, namely the adoption of agile methodologies in Software Engineering and Decision Support Systems discipline were important elements for students to apply personal time management practices that proved to be important in the context of the game; (ii) the group work in large dimension groups (i.e., 6 to 8 students) carried out mainly in the Decision Support Systems discipline proved to be important in developing both time management skills and giving feedback to a project team; and (iii) around 20% of students in the Computer Science course already had an Erasmus + experience, which was important in developing social dynamics.

5.4. RQ4—What Are the Summative and Formative Elements Used in the Assessment of Students' Performance?

The semi-structured interview held at the end of the game was also relevant to systematically identify and expose the summative and formative elements used in the process of assessing student performance. The versatility of FLIGBY allowed the incorporation of both summative and formative assessment components. The main element of summative evaluation is FLIGBY personal report which is a relatively long report that evaluates the player's performance according to the 29 MAP dimensions and lists the key performance indicators (KPIs) of the player's performance. However, despite the unequivocal importance of the summative assessment elements, the main advantage offered by FLIGBY is the introduction of formative elements that give the student important real-time feedback on their performance. These elements are essential to the player's motivation and learning process. As main elements of formative assessment, FLIGBY offers. (i) Mr. FLIGBY, which is a virtual coach that gives advice to the players and praise them if they make good decisions; (ii) the FLOW Profile that allows visualizing the positioning of each player in the eight emotional states of the Flow theory proposed by [47]; (iii) the Flow Meter of the Turul Winery indicates the corporate atmosphere meter which shows the average mood inside the organization; and (iv) Profitability Index which shows how a player's decision impacts Turul Winer's revenue-generating ability. In Figure 4 the following elements are identified: (A) Flow Profile; (B) Flow Meter; and (C) Profitability Index.

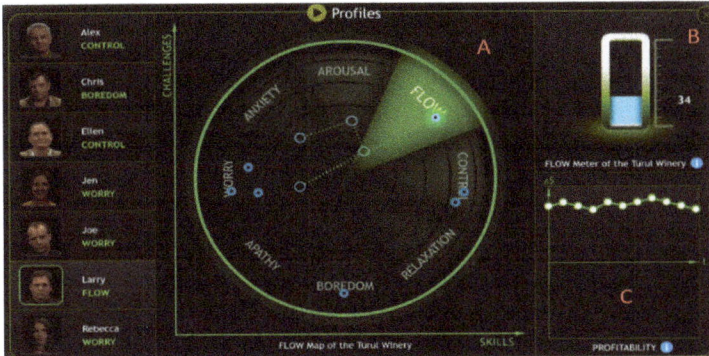

Figure 4. FLIGBY formative feedback (authors' own illustration). (**A**) Flow Profile; (**B**) Flow Meter; and (**C**) Profitability Index.

The inclusion of these four formative elements is fundamental for students to adjust the development of their game processes and strategies as suggested by [44]. This adjustment is performed by the player himself without the intervention of the teacher. Another key aspect offered by the game is the level of challenge that is posed to each player that is adapted according to their actions in the game. As indicated by [45] it is essential that the feedback that is given to the player is personalized according to the different learning styles. Mr. FLIGBY's intervention is important to give indications to the player how he/she can improve his/her performance in the next scenario, and the feedback that was given by Mr. FLIGBY always seeks to be constructive by identifying first of all the correct actions taken by the player, but also indicating the areas in which he/she should improve.

6. Conclusions

Entrepreneurship is a standard of living that includes a set of behaviors and skills that can be developed and applied not only when launching a new business but also to enhance the performance of any job activity. The mission of universities, in addition to preparing students for the job market, is to form critical and aware citizens who can contribute as agents of change in society. In this sense, the inclusion of entrepreneurship in the formal university education system is relevant. FLIGBY presents itself as a technological tool that through a serious game allows the acquisition and development of competencies through a powerful, immersive, and personalized learning space.

The use of FLIGBY in the context of an entrepreneurship course involving students with multidisciplinary skills in management and computer science was received very positively by students. The students mentioned the usefulness of the game in the development of management, leadership and entrepreneurship skills that will be essential to them throughout their academic career and in the labor market. By using FLIGBY, students were able to train their skills in a wide range of domains like gathering information, motivating employees, training their emotional intelligence, and establishing social dynamics in a corporate environment. Furthermore, the use of FLIGBY allowed these students to better understand their skills and explore how they interact with individuals that have very distinct and often conflicting characters. These skills are fundamental in the labor market and are typically not included and addressed in the context of a higher education degree.

FLIGBY offers both summative and formative assessment components. For one side, the FLIGBY personal report is a fundamental element in the analysis of the player's performance in the game considering the 29 MAP dimensions. This report also allows a comparative analysis of the benchmarking obtained by players in each of these dimensions. Despite the importance of summative elements, the potential of FLIGBY is revealed in the inclusion of several formative elements that support and evaluate the player in a non-intrusive way throughout the game. These formative elements are essential in the development of student's skills while they are playing.

This study has unequivocal theoretical and practical potentialities and benefits. From the theoretical point of view, a new informal teaching method based on a serious game is presented that allows students to increase technical skills in the field of management and entrepreneurship and also allows them to develop essential soft-skills (e.g., leadership, group work, emotional intelligence, problem-solving skills, among others) that are crucial when launching a new venture. On the other hand, in practical terms, this study seeks to arouse in other higher education institutions the desire to include serious games as a complementary activity to formal teaching methods in an entrepreneurship course, highlighting these game's relevance. It is intended that this case study serve as a reference for other higher education institutions to adopt serious games in the context of an entrepreneurship course.

As future work, it would be relevant to increase the number of pupils involved in this initiative. Currently, only students attending the 2018/19 school year entrepreneurship discipline have been included in this study, but prospectively it is intended to include students from upcoming editions of this course. Furthermore, and entrepreneurship being a multidisciplinary theme, it is pertinent to involve other engineering and social sciences courses in this discipline. Another aspect to be assessed is the impact of the use of FLIGBY on students' academic performance. To this end, it is important to explore how performance in each of the 29 MAP dimensions has an impact on students' academic success through the use of non-parametric and parametric statistical techniques. Finally, it is also relevant to explore the use of FLIGBY as a coaching tool in which students can practice the development of skills in which they experience greater difficulty.

Author Contributions: Conceptualization: Z.B.; formal analysis: Z.B. and F.A.; investigation: Z.B. and F.A.; methodology: Z.B. and F.A.; project administration: Z.B.; resources: Z.B.; supervision: Z.B.; validation: Z.B.; visualization: F.A.; writing—original draft: F.A.; writing—review and editing: Z.B.

Funding: This research received no external funding.

Conflicts of Interest: The first author is the leader of the "Leadership&Flow" Management Board and was involved in the design and development of FLIGBY. This manuscript presents a case study of a higher education institution that adopts FLIGBY as a serious game in an entrepreneurship course.

References

1. Martin, C.; Iucu, R. Teaching entrepreneurship to educational sciences students. *Procedia Soc. Behav. Sci.* **2014**, *116*, 4397–4400. [CrossRef]
2. Sirelkhatim, F.; Gangi, Y.; Nisar, T. Entrepreneurship education: A systematic literature review of curricula contents and teaching methods. *Cogent Bus. Manag.* **2015**, *2*, 1–11. [CrossRef]
3. Khosla, A.; Gupta, P. Traits of Successful Entrepreneurs. *J. Priv. Equity Summer* **2017**, *20*, 12–15. [CrossRef]
4. Cheung, C.; Au, E. Running a small business by students in a secondary school: Its impact on learning about entrepreneurship. *J. Entrep. Educ.* **2010**, *13*, 45–63.
5. Elmuti, D.; Khoury, G.; Omran, O. Does entrepreneurship education have a role in developing entrepreneurial skills and venture's effectiveness? *J. Entrep. Educ.* **2012**, *15*, 83–98.
6. Lautenschläger, A.; Haase, H. The myth of entrepreneurship education: Seven arguments against teaching business creation at universities. *J. Entrep. Educ.* **2011**, *14*, 147–161.
7. Hamburg, H. Learning Approaches for Entrepreneurship Education. *Adv. Soc. Sci. Res. J.* **2015**, *3*, 228–237. [CrossRef]
8. Welsh, D.; Tullar, W.; Nemati, H. Entrepreneurship education: Process, method, or both? *J. Innov. Knowl.* **2016**, *1*, 125–132. [CrossRef]
9. Askun, B.; Yildirim, N. Insights on Entrepreneurship Education in Public Universities in Turkey: Creating Entrepreneurs or Not? *Procedia Soc. Behav. Sci.* **2011**, *24*, 663–676. [CrossRef]
10. Stamboulis, Y.; Barlas, A. Entrepreneurship education impact on student attitudes. *Int. J. Manag. Educ.* **2014**, *12*, 365–373. [CrossRef]
11. Sousa, M. Entrepreneurship Skills Development in Higher Education Courses for Teams Leaders. *Adm. Sci.* **2018**, *8*, 18. [CrossRef]
12. Potishuk, V.; Kratzer, J. Factors affecting entrepreneurial intentions and entrepreneurial attitudes in higher education. *J. Entrep. Educ.* **2017**, *20*, 25–44.

13. Hytti, U.; O'Gorman, C. What is "enterprise education"? An analysis of the objectives and methods of enterprise education programmes in four countries. *Educ. Train.* **2004**, *46*, 11–23. [CrossRef]
14. Jones, B.; Iredale, N. Enterprise education as pedagogy. *Educ. Train.* **2010**, *52*, 7–19. [CrossRef]
15. Belitski, M.; Heron, K. Expanding entrepreneurship education ecosystems. *J. Manag. Dev.* **2017**, *36*, 163–177. [CrossRef]
16. Laamarti, F.; Eid, M.; El Saddik, A. An Overview of Serious Games. *Int. J. Comput. Games Technol.* **2014**, *2014*. [CrossRef]
17. Bittner, J.; Schipper, J. Motivational effects and age differences of gamification in product advertising. *J. Consum. Mark.* **2014**, *31*, 391–400. [CrossRef]
18. Huotari, K.; Hamari, J. A definition for gamification: Anchoring gamification in the service marketing literature. *Electron. Mark.* **2017**, *27*, 21–31. [CrossRef]
19. Deterding, S.; Dixon, D.; Khaled, R.; Nacke, L. From game design elements to gamefulness: Defining gamification. In Proceedings of the 15th International Academic MindTrek Conference: Envisioning Future Media Environments, Tampere, Finland, 28–30 September 2011; pp. 9–15.
20. Conaway, R.; Garay, M. Gamification and Service Marketing. *Springerplus* **2014**, *3*, 1–11. [CrossRef]
21. Edwards, E.; Lumsden, J.; Rivas, C.; Steed, L.; Edwards, L.; Thiyagarajan, A.; Sohanpal, R.; Caton, H.; Griffiths, C.; Munafò, M.; et al. Gamification for health promotion: Systematic review of behavior change techniques in smartphone apps. *BMJ Open* **2016**, *6*, e012447. [CrossRef]
22. Rowell, L.; Hong, E. Academic Motivation: Concepts, Strategies, and Counseling Approaches. *Prof. Sch. Couns.* **2018**, *16*, 158–171. [CrossRef]
23. Sailer, M.; Hense, J.; Mayr, S.; Mandl, H. How gamification motivates: An experimental study of the effects of specific game design elements on psychological need satisfaction. *Comput. Hum. Behav.* **2017**, *69*, 371–380. [CrossRef]
24. Llorens-Largo, F.; Gallego-Durán, F.; Villagrá-Arnedo, C.; Compañ-Rosique, P.; Satorre-Cuerda, R.; Molina-Carmona, R. Gamification of the Learning Process: Lessons Learned. *IEEE Rev. Iberoam. Tecnol. Aprendiz.* **2016**, *11*, 227–234. [CrossRef]
25. Alvarez, J.; Michaud, L. *Serious Games: Advergaming, Edugaming, Training, and More*; IDATE: Montpellier, France, 2008.
26. Karagiorgas, D.; Niemann, S. Gamification and Game-Based Learning. *J. Educ. Technol. Syst.* **2017**, *45*, 499–519. [CrossRef]
27. Bellotti, F.; Berta, R.; De Gloria, A.; Lavagnino, E.; Dagnino, F.; Ott, M.; Romero, M.; Usart, M.; Mayer, I. Designing a Course for Stimulating Entrepreneurship in Higher Education through Serious Games. *Procedia Comput. Sci.* **2012**, *15*, 174–186. [CrossRef]
28. Almeida, F.; Simões, J. Serious Games in Entrepreneurship Education. In *Encyclopedia of Information Science and Technology*, 4th ed.; Khosrow-Pour, M., Ed.; IGI Global: Hershey, PA, USA, 2018; pp. 800–808.
29. Anghern, A.; Maxwell, K. EagleRacing: Addressing Corporate Collaboration Challenges through an Online Simulation Game. *Innov. J. Online Educ.* **2009**, *5*, 4.
30. Hauge, J.; Bellotti, F.; Berta, R.; Carvalho, M.; De Gloria, A.; Lavagnino, E.; Nadolski, R.; Ott, M. Field assessment of serious games for entrepreneurship in higher education. *J. Converg. Inf. Technol.* **2013**, *8*, 1–12.
31. Fox, J.; Pittaway, L.; Uzuegbunam, I. Simulations in Entrepreneurship Education: Serious Games and Learning Through Play. *Entrep. Educ. Pedagog.* **2018**, *1*, 61–89. [CrossRef]
32. Bellotti, F.; Berta, R.; De Gloria, A.; Lavagnino, E.; Antonaci, A.; Dagnino, F.; Ott, M.; Romero, M.; Usart, M.; Mayer, I. Serious games and the development of an entrepreneurial mindset in higher education engineering students. *Entertain. Comput.* **2014**, *5*, 357–366. [CrossRef]
33. Almeida, F. Experience with Entrepreneurship Learning Using Serious Games. *Cypriot J. Educ. Sci.* **2017**, *12*, 69–80. [CrossRef]
34. Zimmerman, J. Using Business Plans for Teaching Entrepreneurship. *Am. J. Bus. Educ.* **2012**, *5*, 727–742. [CrossRef]
35. Ferreira, F.; Pinheiro, C. Circular Business Plan: Entrepreneurship teaching instrument and development of the entrepreneurial profile. *Gestão Produção* **2018**, *25*, 854–865. [CrossRef]
36. Rosen, A. Why Collaboration Is Essential to Entrepreneurship. Entrepreneur Europe, 2015. Available online: https://www.entrepreneur.com/article/245599 (accessed on 12 April 2019).
37. Ulicsak, M.; Wright, M. *Games in Education: Serious Games*; Futurelab Series: Berkshire, UK, 2010.
38. Mayer, I.; Warmelink, H.; Bekebrede, G. Learning in a game-based virtual environment: A comparative evaluation in higher education. *Eur. J. Eng. Educ.* **2012**, *38*, 85–106. [CrossRef]

39. Sánchez, J. The Impact of an Entrepreneurship Education Program on Entrepreneurial Competences and Intention. *J. Small Bus. Manag.* **2013**, *51*, 447–465. [CrossRef]
40. Mojab, F.; Zaefarian, R.; Azizi, A. Applying Competency based Approach for Entrepreneurship Education. *Procedia Soc. Behav. Sci.* **2011**, *12*, 436–447. [CrossRef]
41. Kissi, E.; Somiah, M.; Ansah, S. Towards Entrepreneurial Learning Competencies: The Perspective of Built Environment Students. *High. Educ. Stud.* **2015**, *5*, 20–30. [CrossRef]
42. Yüksel, H.; Gündüz, N. Formative and Summative Assessment in Higher Education: Opinions and Practices of Instructors. *Eur. J. Educ. Stud.* **2017**, *3*, 336–356. [CrossRef]
43. Black, P.; Harrison, C.; Lee, C.; Marshall, B.; Wiliam, D. *Assessment for Learning: Putting It into Practice*; Open University Press: Berkshire, UK, 2003.
44. Price, V. Exploring effectiveness and rationale of different assessment types. *J. Initial Teach. Inq.* **2015**, *1*, 13–15.
45. Bellotti, F.; Kapralos, B.; Lee, K.; Moreno-Ger, P.; Berta, R. Assessment in and of Serious Games: An Overview. *Adv. Hum. Comput. Interact.* **2013**, *2013*, 136864. [CrossRef]
46. Daoudi, I.; Tranvouez, E.; Chebil, R.; Espinasse, B.; Chaari, W.L. Learners' Assessment and Evaluation in Serious Games: Approaches and Techniques Review. In *Information Systems for Crisis Response and Management in Mediterranean Countries*; Dokas, I., Bellamine-Ben Saoud, N., Dugdale, J., Díaz, P., Eds.; Springer: Berlim, Germany, 2017; Volume 301, pp. 147–153.
47. Csikszentmihalyi, M. *Good Business: Leadership, Flow and the Making of Meaning*; Penguin Books: New York, NY, USA, 2003.
48. Ávila-Pesántez, D.; Rivera, L.; Alban, M. Approaches for Serious Game Design: A Systematic Literature Review. *Comput. Educ. J.* **2017**, *8*, 1–11.
49. Catalano, E.; Luccini, A.; Mortara, M. Guidelines for an effective design of serious games. *Int. J. Serious Games* **2014**, *1*. [CrossRef]
50. Newbery, R.; Lean, J.; Moizer, J. Evaluating the impact of serious games: The effect of gaming on entrepreneurial intent. *Inf. Technol. People* **2016**, *29*, 733–749. [CrossRef]
51. Kafai, Y.; Burke, Q. Constructionist Gaming: Understanding the Benefits of Making Games for Learning. *Educ. Psychol.* **2015**, *50*, 313–334. [CrossRef]
52. Marer, P.; Buzady, Z.; Vecsey, Z. *Missing Link Discovered-Planting Csikszentmihalyi's Flow Theory into Management and Leadership Practice by Using FLIGBY, the Official Flow-Leadership Game*; Aleas Inc.: Los Angeles, CA, USA, 2016.
53. Creswell, J.; Clark, V. *Designing and Conducting Mixed Methods Research*; SAGE Publications: Thousand Oaks, CA, USA, 2017.
54. Almeida, F. Strategies to perform a mixed methods study. *Eur. J. Educ. Stud.* **2018**, *5*, 137–151. [CrossRef]
55. Fowler, F. *Survey Research Methods*; SAGE Publications: Thousand Oaks, CA, USA, 2013.
56. Nardi, P. *Doing Survey Research*; Routledge: Abingdon, UK, 2018.
57. Braun, V.; Clarke, V. Using thematic analysis in psychology. *Qual. Res. Psychol.* **2006**, *3*, 77–101. [CrossRef]
58. Creswell, J.; Poth, C. *Qualitative Inquiry and Research Design: Choosing among Five Approaches*; SAGE Publications: Thousand Oaks, CA, USA, 2017.
59. Wilson, K. Entrepreneurship Education in Europe. 2008. Available online: https://www.oecd:site/innovationstrategy/42961567.pdf (accessed on 5 March 2019).
60. Frese, M.; Gielnik, M. The Psychology of Entrepreneurship. *Annu. Rev. Organ. Psychol. Organ. Behav.* **2014**, *1*, 413–438. [CrossRef]
61. Kouzes, J.; Posner, B. *The Leadership Challenge: How to Make Extraordinary Things Happen in Organizations*; Jossey-Bass: New York, NY, USA, 2017.
62. Razali, S.; Rusiman, M.; Gan, W.; Arbin, N. The Impact of Time Management on Students' Academic Achievement. *J. Phys. Conf. Ser.* **2018**, *995*. [CrossRef]
63. Forsyth, D.; Van Gelderen, M. Time Management for Novice Nascent Entrepreneurs. *Int. J. Entrep. Educ.* **2005**, *3*, 245–252.
64. Keefer, K.; Parker, J.; Saklofske, D. *Emotional Intelligence in Education*; Springer: Berlin, Germany, 2018.

© 2019 by the authors. Licensee MDPI, Basel, Switzerland. This article is an open access article distributed under the terms and conditions of the Creative Commons Attribution (CC BY) license (http://creativecommons.org/licenses/by/4.0/).

Article

Enhancing Fun through Gamification to Improve Engagement in MOOC

Oriol Borrás-Gené [1,*], Margarita Martínez-Núñez [2] and Luis Martín-Fernández [2]

[1] Departamento Ciencias de la Computación, Arquitectura de Computadores, Lenguajes y Sistemas Informáticos y Estadística e Investigación Operativa, LITE Group—Universidad Rey Juan Carlos, 28933 Madrid, Spain
[2] Departamento de Ingeniería de Organización, Administración de Empresas y Estadística, Universidad Politécnica de Madrid, 28031 Madrid, Spain
* Correspondence: oriol.borras@urjc.es

Received: 2 July 2019; Accepted: 25 July 2019; Published: 26 July 2019

Abstract: Massive Open Online Courses (MOOCs), regardless of their topic, are a perfect space to generate, through virtual learning communities associated with them, very valuable resources for their participants and, in general, anyone interested in the topic covered. If in the design of these learning spaces, elements specific to games are added to them, which is known as gamification, we can try to increase the engagement of the student towards the course and, therefore, towards the community. This paper presents an experience of a MOOC of Universidad Rey Juan Carlos (Spain) with a connectivist approach. Aspects such as fun and motivation have been worked on in the design, through the application of gamified activities and the use of elements from social networks, considered as gamification, with the aim of increasing participation and engagement within a Facebook group, used as a community to support the course. We have analyzed aspects such as enjoyment and motivation, the result of which has been active participation and high engagement within the MOOC community in the form of content and especially great interaction, highlighting the existence of continuous activity once the edition of the MOOC is finished, as a consequence of a habit generated in the student.

Keywords: gamification; MOOC; fun; social networks; virtual learning communities

1. Introduction

The generalization of Internet access and globalization have led to a change in aspects such as communication, work, leisure or business [1], requiring every individual to constantly update their knowledge, which is known as lifelong learning. This continuous learning and the existence of Internet requires new models that explain the way in which one learns. Connectivism [2] tries to explain the learning that takes place in the digital age, which is outside of the person, in the Web and the most important aspect will be the ability to create connections with the content and between different individuals. Siemens [2] emphasizes "as knowledge continues to grow and evolve, access to what is needed is more important than what the learner currently possesses".

To respond to these constant needs for training and new forms of learning, universities are opening their content through various initiatives such as the pioneering OpenCourseWare (OCW) [3] or more recently, the Massive Open Online Courses (MOOC) that are offering university educational level since 2008 to thousands of people, with the only requirement of having access to the Internet [4].

Two approaches stand out within the field of MOOC [5,6], on one hand the cMOOC, based on a connectivist approach, in which its pedagogy its pedagogy promotes the social part of the course, try to focus it around the contributions of the students; on the other hand there are the xMOOC, based on content and with features similar to those of a traditional online course, these are the most widely known. There is a trend towards hybrid solutions [7,8], which take advantage of the number of users of xMOOC platforms, with a space in which to manage users and content; and social spaces in which to develop the most connectivist part of cMOOC. The lack of control of information generated on the Web [2] favors the formation of specialized spaces or Virtual Learning Communities (VLC), and specifically in MOOC it is common to find them as spaces where their participants interact and generate conversation and resources. There are different options when choosing a VLC [8], being the most used digital social networks given their nature and which are used by the majority of MOOC participants, this makes them a convenient solution and with the possibility of having some participation.

New methodologies [6] are increasingly used to achieve better results in MOOC, in most cases focusing on one of the drawbacks of this course modality, the large number of dropouts. Gamification is one of the possibilities in which we work to improve these results, this consists [9] in applying elements of game design in contexts that are not games, in this case education. Although there are other interesting aspects in which it is possible to work using gamification, such as the VLC.

The aim of this research is to find out whether, through the application in one MOOC, with connectivism approach, of various gamification techniques, that increase motivation and fun, it is possible to achieve a greater engagement in terms of participation and generate a habit in the use of the VLC. With the purpose of increasing participation in the form of new content and conversations by the various members of the community associated with the MOOC. For the development of this work we have worked with a learning community on Facebook. As a first hypothesis, it has been considered that the very fact of participating in a digital social network such as Facebook and its elements are a way of gamification in itself. It has been analyzed which is the relation that the participants of the virtual community of learning have with the perception of the amusement and its engagement with it, relying on how they affect the characterization of participants according to these concepts, in order to obtain useful patterns of behavior to apply in future editions of the MOOC. This MOOC, from the Rey Juan Carlos University, lasted one month and was given during the month of September 2018, winning the First Award for Educational Innovation MiriadaX.

In the following sections the experience will be developed. In the "Background" section, a theoretical review is proposed about the main aspects of this work. In the following two sections, the design of the MOOC will be described and the main results obtained will be presented. Finally, this paper shows a discussion of the results obtained and ends with conclusions.

2. Background

2.1. Towards a Hybrid MOOC

MOOCs have appeared as a disruptive element in the field of education [10], within the higher education community, allowing thousands of people from all over the world to be trained in a diversity of contents [4,6,11] through Open Educational Resources (OER) and different pedagogical proposals. However, they also have some exceptions and dropout rates of between 60% and 90% of the students enrolled.

Within the philosophy of connectivism as a learning model, the first MOOC [12] appeared in 2008, known as cMOOC, in which it is the students themselves who construct the contents of the course. A new pedagogy that turns the student into the protagonist of the MOOC [8] where both the content and the discussion generated are totally distributed [4] as is the case with the Internet, where learning is produced by the fact of browsing and creates connections. It emphasizes the autonomy of the student, who uses his habitual digital tool to communicate. This supposes numerous tools and technologies associated to the MOOC and it will be necessary to dominate to be able to follow the course, reason why it is required of some advanced digital skills [4,8].

On the other hand, there are the xMOOC, which since its emergence in 2012 has grown rapidly [11] and are the best known and most common, with thousands of students and no limit on the number of enrolled. Their structure is similar to an online course [5], they are organized by modules, objectives, video lectures and have automatic evaluation tests (test) or between peers. Platforms such as Coursera, EdX, Udacity [4] or MiriadaX congregate the majority of users and are composed mainly of universities. It is common to find in these platforms a more traditional model in which the teacher is the expert and the student is simply a consumer of knowledge, as opposed to the cMOOC where roles are changed, and students are given greater protagonism.

Although connectivist MOOCs have an interesting pedagogical model, the main disadvantage is the distributed nature of the students' contributions, it is complex to follow this type of courses due to the number of communication tools used and to bring together their management, compared to the ease of xMOOC models. At the same time, one of the main limitations of xMOOC is the social part [10,13], being relegated to internal and closed forums, not very useful. At the same time, one of the main limitations of the xMOOC is the social part [10,13], being relegated to internal and closed forums for people who are not part of the MOOC. It is usual that they cannot be accessed after the edition is finished, losing all the conversations generated in them. Sometimes it is difficult to follow the conversations in them because of the students' lack of organization in writing their publications. For this reason, new tendencies appear that mix both types of MOOC [7,8], they are called hybrid MOOC or hMOOC, starting from the use of an xMOOC platform where the teaching team creates the structure of the course with the contents and manages the users. On the other hand, one or more external tools are used to complement the social deficiencies of these platforms, generally tools that favor the social part or that allow the application of new methodologies such as gamification [6] through different elements and mechanisms (badges, reputation, points, etc.).

2.2. Virtual Learning Communities

VLCs are spaces or groups of people with interests that focus on a common theme and that allow interaction and dialogue in digital spaces [14,15].

Advantages of VLCs include [6] synchronous and asynchronous communication; permanence of publications for later review; ease of communication between different geographical areas; connection with potential clients or people interested in other people's publications; and unlimited interactivity. The added value of using social networks is that it makes it easier to share content and generate discussion forums [16] with the most organized content, its use being better known by most students. They are increasingly used in the field of education, such as Facebook groups [8,10] that allow the creation of discussion and interaction spaces in a synchronous and asynchronous way.

2.3. Gamification Principles

Games are an element that is increasingly present in today's society, in part due to the rise and possibilities offered by new technologies that allow these games to be applied through numerous devices. There is a tendency to reuse design principles in other contexts that have nothing to do with games such as marketing, health or education. If one extracts those elements of design, key for the success of a game, and they are applied in these new contexts is when one will be able to speak of Gamification [17,18] and especially in education [19]. In the specific case of MOOCs there is a growth in the use of this methodology [20–23] applying gamification to achieve greater engagement of participants.

One of the current problems in education [24] is the lack of commitment and motivation of students, the application of gamification in education has great potential for improving performance, motivation, engagement or fun [22,23] of training activities. This will enable students to improve their skills and knowledge [24] by devoting more time to study and with greater involvement. It should not be forgotten that emotions and feelings [6,25] have a great impact on learning processes, generating greater engagement or on the contrary even reaching frustration within the course, it will be important to take them into account for proper management.

When talking about design it is complex to find a single framework that defines the different existing elements or dimensions [26] without finding a consensus when talking about terms such as mechanics, dynamics or components. Depending on the author, elements such as narrative, challenge, emotion, collaboration, badges, and dots will appear related to these terms.

2.4. Gamification and VLC

The use of digital social networks in education as VLC is becoming more widespread, given the impact it has on learning, with aspects such as motivation, engagement, satisfaction, and interaction [27] enabling online sharing and collaboration. Another aspect to study is the feelings generated when an interaction takes place, through discussions or comments in these networks, which is an important element when related to the activity of users in the MOOC [25].

If one analyzes the inherent elements of social network platforms such as Facebook or Twitter, which promote increased activity on them, one perceives a direct relationship with the gamification elements themselves. For example, the fact of receiving "likes" in a publication is related to aspects such as points and positive feelings related to status or even enjoyment, which Facebook allows to different degrees when indicating that we like a publication. The same happens with the phenomenon of being "followed" in a social network or when our publications are shared. On this aspect authors such as Hansch, Newman, and Schildhauer [26] propose a framework when talking about gamification design, in which they include another aspect that features prominently in online learning and knowledge sharing: the social/interactive dimension.

Using these elements of social networks in VLC, we are able to apply gamification without the need for other technologies, not only for the fact of working the social dimension, but also for the use of other dynamics. Specifically, the digital social network Facebook has a tool called "Groups" that allows creation of workspaces, with different privacy settings. This platform is committed to the educational use of these groups and within these, elements closely linked to gamification stand out, such as these:

- Progress bar, associated with the additional functionality of the groups, "Units", which serve to organize the collection and ordering of group publications, managing the chaos that often occurs with so much published information. Members of group are able to mark the publications of each unit seen, checking in the bar its level of progress.
- Digital badges that Facebook itself automatically associates according to the degree or type of participation of members in the group. These, when obtained, are associated with the name of the member who has obtained it, appearing in all the publications made within group. Facebook Groups currently offers 10 different badges [28].
- Different degrees of "likes" (I like it, I love it, it saddens me, etc.), comments and reshares. Although these options are not specific to groups, they are interesting as gamification elements.

3. Materials and Methods

3.1. MOOC

The MOOC analyzed, "Empower yourself with social networks" from Universidad Rey Juan Carlos, was offered in 2018 on the Spanish platform MiriadaX for one month (29 August to 25 September). It obtained First Award for Educational Innovation MiriadaX from the same platform in 2018 [29]. Over 4 weeks and 5 modules, the MOOC works on aspects such as personal branding on the Internet. Each module consists of a series of lessons that work on specific topics through video, text, links of interest, and exercises. The first week introduces the course learning guide (module 0) and contextualizes the course theme (module 1) emphasizing aspects such as professional digital identity. The next two weeks is where they see different social networks, from those more general as Facebook or Twitter (module 2) to others more specialized, such as LinkedIn (module 3). The fourth and final module works tools such as video, web analytics, and the concept of Personal Branding as the final result of the MOOC. All contents of the MOOC were accessible from the first day, except exams of each module which were opened one per week. In this way the students could review the material at their own pace, although through e-mail communications was set a rhythm and order, for those who needed a more guided modality.

The proposed design is based on a hybrid philosophy [7] called gcMOOC, already used in other works by the authors [21], which takes advantage of the infrastructure already created, such as the MiriadaX platform (xMOOC), and which has more than 6 million registered users [30], to offer a cMOOC approach through external elements [8]. The objective is to achieve greater student involvement and increased interaction in the MOOC in the form of new resources, thus following the connectivist principles, where priority is given to the content provided by those participating in the MOOC. For this, it was necessary to use a virtual learning community in which to generate a connectivist part. The key to the proposed model is the use of gamification, especially taking advantage of the social part of it, from a digital social network such as Facebook, and its own gamified elements that seek greater engagement.

Therefore, the MOOC is divided into two spaces, the platform where the course is, and the VLC where the activity is generated. In the following sections the proposed model will be analyzed: both the virtual learning community created and those gamification elements introduced.

3.2. Virtual Learning Community (Facebook Group)

In order to encourage participation, a virtual learning community was created on the Facebook platform, specifically a public group. The advantage of groups is that they organize publications on a wall where any member of the group can publish content appearing in the first place each publication, unlike Facebook pages, where only appear on the wall the publications of the administrators. Groups have very complete statistics for the teacher and also have interesting features for education: units, badges, publications with styles (bold, italics, numbering, etc.), hashtags, different roles of members and systems for approval of publications or searches within the community.

The VLC created for the MOOC is totally public and presents a set of rules of conduct previously defined in a corresponding section. In addition, in the MOOC, there is a lesson within module 0 that links to the community and explains how it works through a video tutorial. The link to the community is constantly quoted in e-mails sent to students and in various MOOC lessons.

The "Units" tool of the group has been used, which allows the teacher to organize those publications that he considers appropriate within the units. In the case of the MOOC they were used as tips or pieces of advice, to get the most profit from each module. Four units or tips were created, within which the teacher organized all the publications of the team, to avoid that these were lost in the time line of the group between so much publication of the participants. The units allow to make them obligatory, in such a way that they appear to the student as incomplete in a progress bar that as the reading seeing will be completed. Figure 1 shows on the left a unit, one can see the bar that indicates graphically and numerically the number of publications of the unit viewed and an example of publication, at the end the option can be seen "done" that the user can mark once viewed; to the right is the information message that appears to each user after finishing a unit and below the general bar that every user sees in the group, indicating the number of completed units.

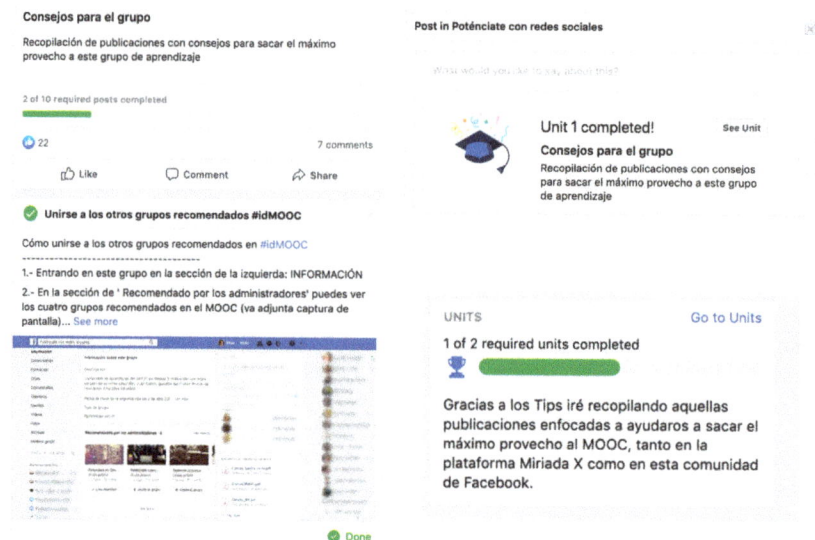

Figure 1. Screenshots of units in Facebook Groups.

3.3. Applied Gamification

The xMOOC models are the most massified and therefore generalized. In these courses most users choose to make use of content and individual learning; few go beyond and work a more social and connected part. Hybrid approaches, such as this model, with external communities where most MOOC activity takes place. The drawback is, on the part of MOOC students, ignorance of their existence, lack of interest or digital skills, this will mean that they do not access these spaces or do not make the most of them.

In this work, through gamification, we seek to motivate and generate new habits in the student to see the usefulness of the VLC and become accustomed to access and work in it. To this end, the two spaces have been worked on: the MOOC and the VLC.

3.3.1. MOOC and VLC gamification

In the first place, it has been decided to use challenges that involve the student through voluntary activities in which they can put into practice the theoretical concepts set out in the MOOC and, on the other hand, encourage participation in the VLC sharing the results. They were divided into two types:

- Asynchronous activities: A total of 12 exercises distributed within the lessons, in which the student can put into practice what he has learned in the lesson at his own pace, and which encourage him to share the result in the form of a publication in the virtual learning community, thus seeking a flow of content from the MOOC to the VLC [8,31]. They were proposed activities such as creating a profile in the social networks explained and sharing it in the VLC, or looking for additional information about the contents, also sharing it.
- Synchronous activities: 4 interactive live activities were proposed, in form of a broadcast via YouTube video platform, using "Hangouts on-air" [32] live videoconferencing option. Any student with the link could view the live event and participate in the chat on the right side of the video.

Within these synchronous activities, the novelty with respect to other initiatives of the team has been the design, seeking a very active participation and protagonism of the student within the activity and to feel part of this. This is the summary of the events:

1. "Building digital identity": In this event, the professor proposed a collaborative analysis of what actions are carried out on the Internet that leave their mark on the digital identity of any person. We used the chat of the YouTube retransmission through which to participate live; the professor was working on the ideas he was reading written in the chat.
2. "What about us on Internet": In this case was made use of the Kahoot! platform of questions, which allows to use it to apply gamification, given its format of a contest. It was used live, allowing anyone to participate who entered the platform, via web or mobile app, and wrote the specific code provided during the live event for the space created. Through a survey, the activity proposed to participants to make different Internet searches about their name and surname, and to see what information appeared from them, analyzing the results the teacher and contributing ideas.
3. "Interview with...": The last two events had the same configuration, they were two interviews, one per week, to experts in the subject of each week module. The event was designed for maximum participation and protagonism of the students. Four days before the event, a link was sent to a space in the Sli.do questions and answers platform, where any MOOC student could write a question on his or her behalf or anonymously, leaving the question published for anyone to see and vote for. The most voted questions were those that were asked to each one of the interviewees.

On the other hand, the use of a digital social network as a support to the VLC and from the point of view of the applied gamification, not only allows the use of a more social and collaborative dimension, it also offers its own elements (likes, units, shares, and badges) that could be adopted within the gamified design of the MOOC, which tend to be unnoticed in the literature. These kinds of elements encourage dynamics such as socialization, participation, debate, status or progress. Figure 2 shows an example of how badges associated with a profile would look (admin and visual storyteller).

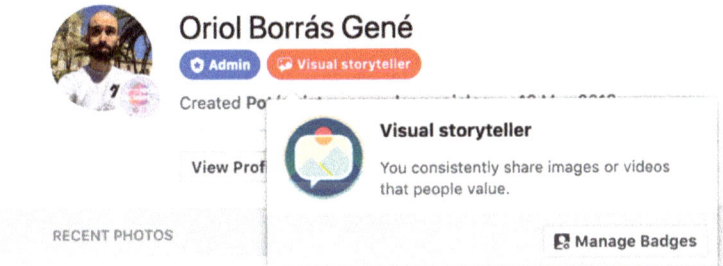

Figure 2. Digital badges in Facebook Groups.

3.3.2. Gamified Model

After reviewing the different actions in the two main spaces of the MOOC, Figure 3 shows the resulting design. At the top is the hybrid MOOC model that uses a VLC to support the most connectivist and social part. In order to increase engagement, gamification is applied in this work, the result being checked in the lower part of the figure.

Informatics **2019**, *6*, 28

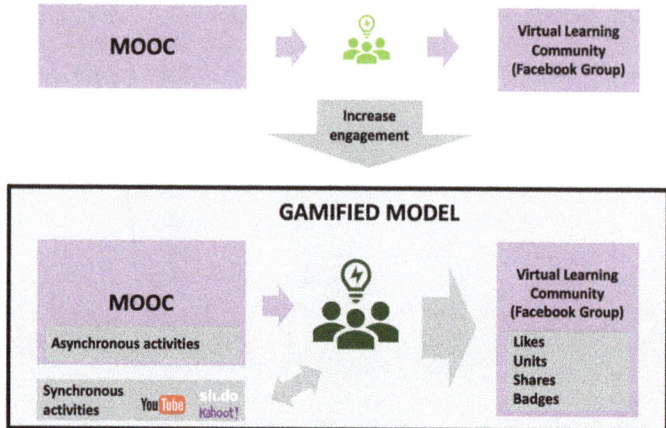

Figure 3. Gamified model (Massive Open Online Courses (MOOC) and Virtual Learning Communities (VLC)).

The most structured and organized part is associated with the MOOC (left) and is where both asynchronous activities are proposed, within the lessons, and synchronous activities, at different moments of the edition, which make use of external tools. The result seeks to generate an increase in publications and greater interaction within the virtual community of learning, on the right of the figure is shown this community and what gamified elements within it also encourage motivation to participate and interact within.

Table 1 summarizes the features, tools, and platforms used to apply the gamification in MOOC edition (Figure 3).

Table 1. Relationship between gamification activities and technologies used.

	Technologies	
Asynchronous Activities	MiriadaX (MOOC platform)	
Synchronous Activities	Kahoot!	
	Sli.do	
Virtual Learning Communities	Facebook Group	Likes
		Shares
		Units
		Badges

3.4. Data Collection

In order to obtain data, we used the statistics provided by the platforms on which MOOC was designed and all its activities:

- MiriadaX allows you to know the number of students enrolled, how many have started and how many have finished.
- Facebook Group offers very detailed statistics, showing information such as active members in the group, number of publications detailed by date, or visits and reactions to publications.
- YouTube allows to know in a very complete way the visualizations of the videos and of the live broadcasts.
- Sli.do and Kahoot! also provide statistics of participation, although they are very basic, for the purpose of the experience have been enough.

At the end of the MOOC, a satisfaction survey was sent to all enrolled, based on the validated SEEQ survey [33], already used in other experiences of the same team. Specific questions were added related to specific aspects to be measured on the MOOC the VLC, such as the tools used or the perception of learning and fun. A total of 63 Likert type questions with a range from 1 to 5 (Strongly disagree to Strongly agree) were posed. In some questions the option "I don't know him" was added to make include this possibility for the specific case of tools used.

4. Results

The research presented in this article works with the data extracted from student activity in an MOOC designed and delivered by the teaching team over the 28 days, comparing them with a satisfaction survey sent after completion.

Table 2 reflects the overall results of the completion of MOOC and the virtual learning community on the last day of the edition (25/09/2018). The number of completions corresponds to students who passed at least 75% of the assessment activities and video lectures viewed. The percentages of finalization have been calculated in relation to those who registered and also to those who started the MOOC, this number is more significant, bearing in mind that the platform considers as initiated any user who has accessed at least once during the course period, as opposed to the 3980 registered users who did not enter even once during the edition, and therefore did not see any video or content.

Table 2. Results of MOOC's participation in the MiriadaX platform and Facebook group.

MOOC (MiriadaX)				
Enrolled	Started	Finished (100%)	% (Enrolled)	% (Started)
8532	4552	1359	15.9%	29.8%
		Finished (75%)	% (Enrolled)	% (Started)
		1655	19.4%	36.3%
VLC (Facebook Group)				
Members		Publications	Comments	Reactions
2540		1980	2746	13,333

As for the VLC, this table shows the total values after completing the course, highlighting a large interaction of users with respect to publications with comments and reactions. The latter refer to the Facebook "likes" of both publications and comments, taking into account the different likes options offered by the social network.

All events have been supported by Hangouts on air from YouTube, broadcast live. Table 3 shows a summary of the activity of these four events. This table shows, for each event, the Youtube information, in time ranges, from the day of the event, until the end of the edition and until the day before the second edition of the MOOC began. The values obtained are from the day of recording. The average duration was 30 min, and you can see the fall of users who saw them, and therefore engagement in the MOOC, as the edition progresses.

4.1. Live Events

Throughout the MOOC different types of activities were proposed, as seen above. The results of the synchronous gamification activities and the tools used to interact with the students will be analyzed below.

Table 3. Live events summary.

		Live Events (Synchronous Activities)			
		Building Digital Identity	What about Us on Internet	Interview with... (1)	Interview with... (2)
YouTube					
Streaming duration (min:sec)		32:25	26:05	33:01	27:41
Event day	Day	04/09/2018	11/09/2018	18/09/2018	25/08/2018
	Views	798	803	270	381
	Peak concurrents	182	157	83	78
	Unique viewers	599	534	214	319
	Likes	71	41	59	48
	Shares	19	8	6	7
	Chat messages	373	-	-	-
Until end of MOOC (25/09/2018)	Period	22 days	15 days	7 days	1 day
	Views	2834	1333	608	515
	Unique viewers	1963	886	444	412
	Likes	110	52	70	53
	Shares	83	12	14	8
After finishing MOOC 26/09/2018 to 01/04/2019	Views	3290	1407	672	664
	Likes	112	55	70	57
	Shares	96	12	10	8
	Comments	46	32	5	3
Kahoot!	Participants	-	88	-	-
Sli.do	Participants	-	-	49	41
	Questions	-	-	50	36
	Likes	-	-	102	52

As for the interaction generated in the events, different elements were used to allow the participation of the spectators. For the first one we used YouTube's own chat, with a total of 373 live conversations. Kahoot!, a very interesting application used in education to gamify, got a total of 88 participants, 10% of the viewers, who answered the 8 questions posed. The last two events asked for a previous participation through Sli.do. In the 4 days prior to each event MOOC students proposed 50 and 36 questions respectively for the interviewees, and these obtained 102 and 52 likes. Sli.do allows anonymous questions to be published or the name to be written, and only 36% were anonymous at both events.

4.2. Participation in the VLC

In this section the specific results obtained in the Facebook community will be analyzed. Figure 4 shows the generated activity in a graphical way. At the top of the graph you can see the evolution of the publications (red), this shows a downward trend as the end of MOOC edition approaches, versus the interaction (comments and reactions) between members that has been more constant. It emphasizes that the number of active users remains constant throughout the edition. The lower part of Figure 4 shows the evolution of community members, which has been growing even at the end of the edition. During the whole edition, both students who started at different times, even at the last minute, and new enrollees were incorporated into the community, as the platform allows registration until the end of the edition.

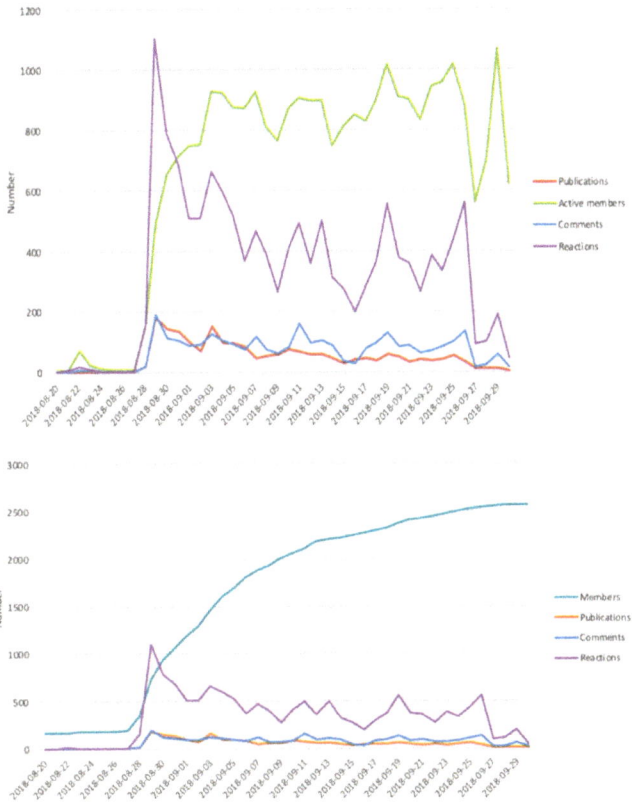

Figure 4. VLC publications (**top**) and members evolution (**bottom**).

In order to check whether a compromise had really been achieved and to create a habit in the use of VLC, independently of MOOC, Facebook group was analyzed as soon as the edition of MOOC had been completed, without any kind of community dynamization or emails encouraging participation. Figure 5 shows the result of the activity between 26 September, 2018 and 1 April, 2019, the day before the start of the second edition of the MOOC. The result was an increase of 425 new members, 387 publications, 605 comments, and 2755 reactions. An average of 53.4 active members per day, albeit with a large deviation (196.4).

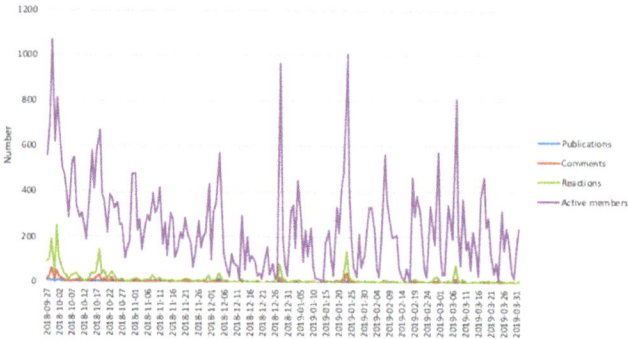

Figure 5. VLC activity after MOOC edition.

4.3. Satisfaction Survey

After completing MOOC, a survey was sent to find out the assessment of students, answered by a total of 480 individuals, that is, 10.5% of those who started the course. Although the survey had a total of 63 questions, the seven most representative questions in relation to this work, summarized in Table 4, have been selected.

Table 4. Selected questions from the satisfaction survey for the study.

Q1. Enjoyment	I had fun with the MOOC
Q2. Gamification	I felt like I was in a game in the MOOC
Q3. Participation	I have shared in the Facebook community the proposed activities of the course
Q4. Participation (a, b, c, d e)	Participation in the Facebook group (a) reading, (b) publishing, (c) commenting to others, (d) likes and (e) sharing publications
Q5. Motivation (a, b, c)	It encourages me, when I publish something, that my publications: (a) have comments, (b) are shared and (c) have likes
Q6. Engagement	I have felt part of the community
Q7. Completion	I've passed the MOOC

Firstly, the students' perception of three aspects related to the MOOC was analyzed: enjoyment and gamification; directly related to questions 1 and 2 respectively. Figure 6 shows each of the three aspects related to question 7 on whether the MOOC was completed at 100%, 75%, or not.

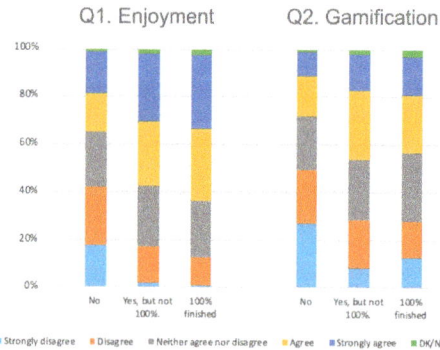

Figure 6. Relationship between learning, enjoyment, gamification and completion of MOOC.

The aspect of enjoyment or fun again highlights that those who have had the most fun have been those who have finished at 100% with 61.6%, and 75% with 55.6%. In both cases, those students who have completed the MOOC consider that they have had fun, an aspect that was sought to be promoted. As for the applied gamification, they were asked if they had felt like in a game, in this case the value as expected decreases, since a complete gamification was not applied, for example to MiriadaX platform itself, even so more than 40% of those who finished the MOOC, have had this perception.

Continuing with these two aspects, Figure 7 shows the relationship, in this case, with respect to the age of the students.

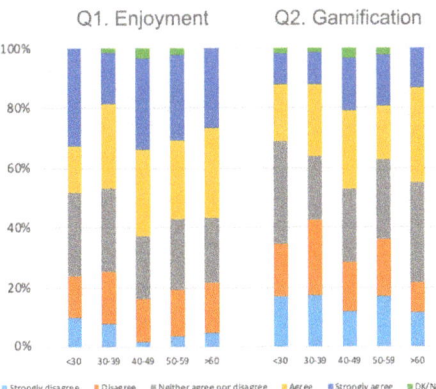

Figure 7. Relationship between learning, enjoyment, gamification and age.

For fun, the values are around 45.3% of the 30–39 age group and 48.3% of those under 30; and the maximum values of 59.5% for the 40–49 age group and 56.7% again for those over 60. The other aspect, most related to gamification, only the age range between 40–49 with 43.9% and over 60 with 45% have felt in a game. As expected, those under 30 were the most skeptical with 29.3%.

Secondly, the aspects related to the VLC on Facebook and the MOOC have been analyzed. On the one hand to check if the applied gamification has increased the participation and on the other hand to check if its elements have really achieved a real commitment of its members with the community and can be considered as motivating elements.

Figure 8 shows the answers given to the degree of participation in the Facebook group using its different elements.

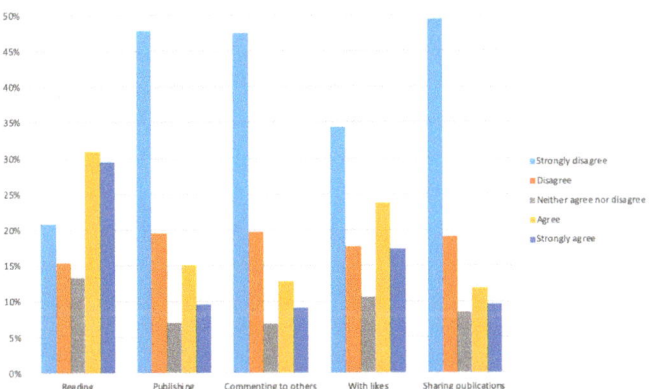

Figure 8. Participation in Facebook group: reading, publishing, commenting, likes, and sharing (Q4).

In general, members of group have read the publications of others, drastically reducing the rest of participative actions, where almost 50% of those asked not shared publications, commented on others or published; although it stands out that between 20% and 24% have actively carried out these actions. The second most common interaction was the use of "likes", with just over 41%.

The following graph (Figure 9) analyzes the results related to motivation in the VLC, from question 5, in order to know how it affects the interaction of others on the subject's participation in the community when publishing.

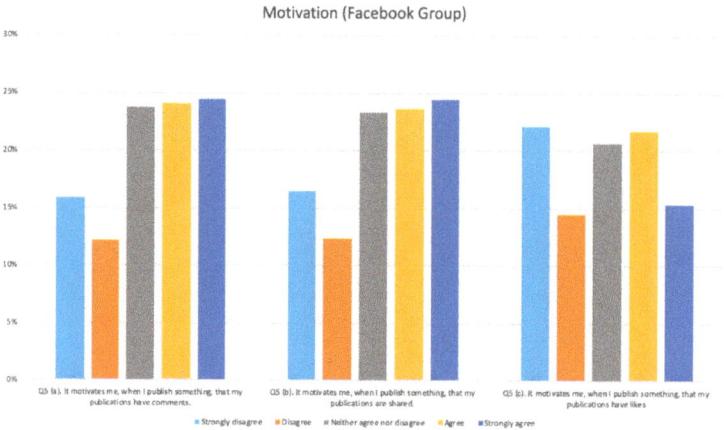

Figure 9. Motivation in Facebook Group (Q5).

Answers to this fifth question reveal that, in general, members of the group are motivated when someone comments on their publications with 48.3%, that they share with 47.9% and that they get "likes" only 37%.

In order to obtain an overview of the results of the previous graphs, Figure 10 proposes a more complete analysis of question 5 from the point of view of the ages of the participants. In this way, the aim is to understand in depth the motivation for a user that others interact with their publications.

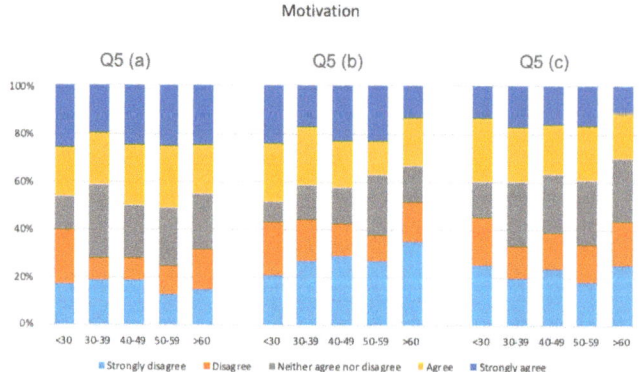

Figure 10. Motivation in Facebook Group (Q5) age analysis.

After the results shown in the graph, it is the participants between 50 and 59 who are most motivated to share and comment on their publications around 51%, being in turn the least important and motivators seem to them the "likes" with only 36.5% interest in them, along with those over 60 who are even less interested with 30%. In general, the rest of ages are motivated with more than 41%

to comment and share with them, being the age groups under 39 who have more attraction for "likes" with 40%.

Finally, fun has been studied as a key element to generate an engagement in the MOOC and, as a consequence, a greater participation in the learning community. Table 5 shows a comparison between the student's perception of fun when performing the MOOC (Q1) and the questions that refer to the feeling of being in a game (Q2), being part of the community (Q6), and about participation in Facebook (Q3 and Q4 (d)). The most representative values of the results have been highlighted in bold, which will be reviewed below.

Table 5. Relationship between gamification, community engagement and fun.

		Q1. I had Fun with the MOOC				
		Strongly Disagree	Disagree	Neither Agree nor Disagree	Agree	Strongly Agree
Q2. I felt like I was in a game in the MOOC	Strongly Disagree	16.82%	39.19%	10.53%	4.72%	2.26%
	Disagree	20.56%	47.30%	30.70%	9.45%	1.50%
	Neither Agree nor Disagree	25.23%	12.16%	50.88%	36.22%	12.03%
	Agree	17.76%	1.35%	7.02%	**48.03%**	**31.58%**
	Strongly Agree	18.69%	0.00%	0.88%	1.57%	**50.38%**
Q6. I have felt part of the community	Strongly Disagree	73.91%	14.86%	1.75%	0.00%	0.75%
	Disagree	26.09%	44.59%	21.05%	5.51%	1.50%
	Neither Agree nor Disagree	0.00%	37.84%	56.14%	23.62%	1.50%
	Agree	0.00%	1.35%	16.67%	**47.24%**	**24.06%**
	Strongly Agree	0.00%	1.35%	4.39%	**22.05%**	**67.67%**
Q3. I have shared in the Facebook community the proposed activities of the course	Strongly Disagree	**86.96%**	51.35%	34.21%	36.22%	22.56%
	Disagree	8.70%	21.62%	22.81%	23.62%	14.29%
	Neither Agree nor Disagree	0.00%	17.57%	24.56%	21.26%	22.56%
	Agree	4.35%	6.76%	13.16%	**10.24%**	**20.30%**
	Strongly Agree	0.00%	2.70%	5.26%	8.66%	**20.30%**
Q4 (d). Participation in the Facebook group [Likes]	Strongly Disagree	73.91%	47.30%	32.46%	37.01%	21.05%
	Disagree	13.04%	22.97%	21.05%	19.69%	12.03%
	Neither Agree nor Disagree	4.35%	16.22%	16.67%	9.45%	4.51%
	Agree	2.17%	7.43%	17.54%	23.62%	28.20%
	Strongly Agree	6.52%	6.08%	12.28%	10.24%	**34.21%**

It is verified that those participants who have had more sensation of being in a game (Agree or Strongly agree in Q2) are those who have had more fun (Agree or Strongly agree in Q1), the same relation that exists with the feeling of being part of the community (Agree or Strongly agree in Q6). About sharing publications, the 86.96% that have not shared publications (Strongly disagree in Q3) are those that have less fun in the MOOC (Strongly disagree in Q1). The surveyed participants that more actively share their publications (Agree or Strongly agree in Q3) are those who have had more fun (Agree or Strongly agree in Q1). At last, those participants who have put more likes (Strongly agree in Q4), have also had fun (Strongly agree in Q1), being these close to 35%. Almost 74% of those who did not put likes (Strongly disagree in Q4) did not have a perception of fun with the MOOC (Strongly disagree in Q1).

5. Discussion and Conclusions

In general, MOOCs are known and criticized for the low student numbers that complete them. Although this value has not been demonstrated to be dependent on the quality of the contents [1], neither is the objective of this research to improve it, if it is true that it can be used to get an idea of the degree of commitment to the course on the part of the students. It is common in MOOCs to find a high drop-out, around 90%–95% [5], compared to the results obtained in this work with 19.4%, which double the most common values. Specifically, 15.9% saw all the contents and surpassed all the activities to 100%.

This research has analyzed the convenience of applying gamification actions in a hybrid MOOC model, where special importance has been given to the social part, promoting those characteristics of a cMOOC, to increase engagement, fun and create a habit when using the VLC, in order to generate new content and conversation.

Two types of activities were proposed in the design, giving rise to a particularly active community with almost 2000 publications during the 28 days of publication. There has been a great deal of interaction between community members with over 2500 comments and 13,000 interactions. Asynchronous activities, associated with content within the MOOC, in which the student was encouraged to publish the results in the community, as well as in other team initiatives in previous MOOCs [21,32]. Synchronous activities, through live events, were quite seen, although with the classic curve of fall in the participation of the MOOC [7] and in them the audience participated, even though they had to make use of different technologies like Kahoot! or Sli.do, we must not forget that the type of MOOC student is very heterogeneous and does not always have advanced digital competences [4,8]. Participation is remarkable in these activities, to a greater or lesser degree, with 1963 visitors who during the MOOC entered the first event, 43.1% of those who started the course, up to 9% of those who saw the last event, as shown in Table 3. This participation is quite high, bearing in mind also that these were voluntary activities and meant that the student had access to a space other than the MOOC platform. Participation has been quite high, bearing in mind that these were voluntary activities and meant that the student had access to a space other than the MOOC platform. Within these events, except the first where the interaction was simple, through the chat of the video itself on YouTube, the rest had mainly viewers in front of students who participated actively entering platforms where the activity was performed as Kahoot! or Sli.do. It should not be forgotten that these types of initiatives require some digital skills, and although its use was a novelty with respect to other MOOCs, its use was minority.

Figure 5 shows that there has been activity, once the edition of the MOOC is finished, both in publications and in interactions, therefore, it has been possible to create a habit and for the members to see the usefulness of participating in this type of communities. Gamification promotes socially desirable learning behaviors [27], in this case the use of the social network itself as a learning space in which to share content.

Based on the satisfaction survey, the students' perception of the sensation of being in a game and having fun participating in the MOOC and the VLC has been analyzed. Figure 6 reveals that those who have had the most fun are those who have finished the MOOC at 100% with 61.6% and 75% with more than 50%, fun is an important element when promoting engagement [23] and to take into account when designing a gamified model [21]. Regardless of the ending, if we focus on age, and reviewing the two graphs globally, we can see that in general the groups that have had the most fun have been from the age of 40. And more specifically, only those over 40 have really had some perception of being in a game, as opposed to those under 30, which as expected have a different concept of game, familiar with mobile applications or real videogames.

From the point of view of motivation, another key element to achieve greater engagement in the model presented, participation in the Facebook group (Figure 8) reveals how students have worked within the community. Highlighting the reading and decreasing the rest of the actions within Facebook, being giving "likes" to others, over 41%, what has been used most, followed by commenting and sharing publications. But, if we change the focus and look at the motivation to receive this type of recognition in the publications or interactions that the student makes, the results change, (Figure 9) with almost 50% to those who do motivate the fact that they are commented or shared. These feelings generated by the interaction (comments, likes, etc.) have a relation with the activity of the students in the MOOC [25]. Within this point of view, it is interesting how the participants are predisposed to give "likes" but only 37% are motivated to receive them. Looking at the age analysis in Figure 10, as expected, the younger generations under 39 are the most motivated to receive likes (41%), to share their publications (40%–45%) and to comment (46%–48%) on generations more habituated to the use of social networks [34], as opposed to those over 60, of which only 30% were interested in receiving likes. The younger generations, who have shown a greater interest in these elements, are more susceptible to what is known as gameful experience also in contexts that are not games [35], having a predisposition to greater engagement when applying these proposals.

Analyzing this type of elements, typical of social networks such as likes, comments or the possibility of sharing other people's publications, offer a "visible status" and a "social engagement" [36]. The ability to follow each other's learning progress, upvoting other peers [26] can harness these motivational aspects of gamification to stimulate participation and engagement with learning contents and with other participants [27], as has been shown in this work. The use of VLC is more and more common in education [27], making use of social networks such as VLC it will be possible, in an indirect way, to take advantage of the use of these own gamification characteristics [26].

The findings shown in Table 5 reinforce this relationship between fun and social part, from feeling in a community where those participants who have had the most fun have also felt a greater degree of community with 67%. On the other hand, no direct relationship has been found between the degree of fun and the sharing of MOOC activities in the community. Finally, there is a certain relationship among the participants who have used "likes" the most and those who have had the most fun with 45%, seeming to them an attractive element, and we consider that it could be associated with the gamification points and the sensation of happiness that it produces, as other authors [26,27].

It is clear that those participants who have had the greatest feeling of being in a game (Q2) are those who have had the most fun, the same relationship that exists with the feeling of being part of the community. Of those surveyed, 23.13% took an active part in sharing their publications, of which around 30% thought it was fun, as opposed to those who did not share anything and did not perceive any fun in the MOOC. Finally, those participants who liked it the most also had fun, being close to 45%. Almost 74% of those who did not put likes did not have a perception of fun with the MOOC. Participants that agreed or strongly agreed with having fun with the MOOC shows higher levels of gamification (Q2), community feeling (Q6), and participation (Q4). However, there is no direct relation between having fun and sharing MOOC activities in Facebook. Therefore, the participation of students in sharing their own activities is not enhanced through having fun. Other actions to motivate sharing activities should be implemented.

As a main conclusion, through the use of gamification, by using two types of activities (synchronous and asynchronous) from a MOOC, and the elements of digital social networks themselves (Facebook), we have achieved greater engagement and generation of content for the VLC. It has been proven that users have had fun and active participation in the community, creating a habit beyond the edition of the MOOC. Fun can be key to achieving greater engagement both in the course, with a higher degree of completion, and in the community.

This increase in motivation to participate in a social network is thanks to attractive elements that can be used as gamification by their own nature, as opposed to more restrained spaces such as online course forums. It has been proposed that members of the community receive an incentive to publish, with this type of interest sample towards their contributions.

With respect to previous experiences, we have worked more on the interaction with the student, to make it more active, incorporating tools such as Kahoot! or Sli.do in live events, although with an extra difficulty for certain users due to lack of digital knowledge, as well as inconvenience when having to use more tools.

The next line of research will seek to deepen the students' perception of the usefulness of the content generated, increase the number of interactive activities in which to apply gamification, incorporating new digital tools. We will also work on the design of activities that are more fun and attractive for students.

Author Contributions: Conceptualization, O.B.-G.; Methodology, O.B.-G.; Validation, O.B.-G., M.M.-N. and L.M.-F.; Formal Analysis, M.M.-N.; Investigation, O.B.-G., M.M.-N. and L.M.-F.; Resources, O.B.-G.; Data Curation, O.B.-G.; Writing—Original Draft Preparation, O.B.-G.; Writing—Review & Editing, O.B.-G. and L.M.-F.; Supervision, O.B.-G.; Project Administration, O.B.-G.

Funding: This work was supported by research grants iProg of MINECO (ref. TIN2015-66731-C2-1-R) and has funded by the Madrid Regional Government, through the project e-Madrid-CM (P2018/TCS-4307). The e-Madrid-CM been co- project is also co-financed by the Structural Funds (FSE and FEDER).

Acknowledgments: This work has counted on the collaboration of the URJC Online team for the design, recording of videos and the management of the MOOC; also, to thank the team of the LITE group.

Conflicts of Interest: The authors declare no conflict of interest.

References

1. Stracke, C.M. The Quality of MOOCs: How to improve the design of open education and online courses for learners? In Proceedings of the 4th International Conference on Learning and Collaboration Technologies, Vancouver, BC, Canada, 9–14 July 2017; Springer: Cham, Switzerland, 2017; pp. 285–293.
2. Siemens, G. Connectivism: A learning theory for the digital age. *Int. J. Instr. Technol. Dis. Learn.* **2005**, *2*, 1–8.
3. Martinez, S. OCW (OpenCourseWare) and MOOC (open course Where?). In Proceedings of the OpenCourseWare Consortium Global, Ljubljana, Slovenia, 23–25 April 2014.
4. Siemens, G. Massive Open Online Courses: Innovation in Education? Available online: https://oerknowledgecloud.org/sites/oerknowledgecloud.org/files/pub_PS_OER-IRP_CH1.pdf (accessed on 2 July 2019).
5. Sein-Echaluce, M.L.; Fidalgo-Blanco, Á.; García-Peñalvo, F.J. Adaptive and cooperative model of knowledge management in MOOCs. In Proceedings of the 4th International Conference on Learning and Collaboration Technologies, Vancouver, BC, Canada, 9–14 July 2017; Springer: Cham, Switzerland, 2017; pp. 273–284.
6. Leony, D.; Merino, P.J.M.; Valiente, J.A.R.; Pardo, A.; Kloos, C.D. Detection and Evaluation of Emotions in Massive Open Online Courses. *J. UCS* **2015**, *21*, 638–655.
7. Fidalgo-Blanco, Á.; Sein-Echaluce, M.L.; García-Peñalvo, F.J. From massive access to cooperation: Lessons learned and proven results of a hybrid xMOOC/cMOOC pedagogical approach to MOOCs. *Int. J. Educ. Technol. High. Educ.* **2016**, *13*, 24. [CrossRef]
8. Borrás-Gené, O. Empowering MOOC Participants: Dynamic Content Adaptation Through External Tools. In Proceedings of the European MOOCs Stakeholders Summit 2019, Naples, Italy, 20–22 May 2019; Springer: Cham, Switzerland, 2019; pp. 121–130.
9. Groh, F. Gamification: State of the Art Definition and Utilization. Available online: http://hubscher.org/roland/courses/hf765/readings/Groh_2012.pdf (accessed on 2 July 2019).

10. Pelet, J.E.; Pratt, M.A.; Fauvy, S. MOOCs and the integration of social media and curation tools in e-learning. In Proceedings of the International Workshop on Learning Technology for Education in Cloud, Maribor, Slovenia, 24–28 August 2015; Springer: Cham, Switzerland, 2015; pp. 43–53.
11. Daniel, J.; Cano, E.V.; Cervera, M.G. The future of MOOCs: Adaptive learning or business model? *Int. J. Educ. Technol. High. Educ.* **2015**, *12*, 64–73. [CrossRef]
12. Rodriguez, O. The concept of openness behind c and x-MOOCs (Massive Open Online Courses). *Open Praxis.* **2013**, *5*, 67–73. [CrossRef]
13. Ramírez-Donoso, L.; Pérez-Sanagustín, M.; Neyem, A.; Rojas-Riethmuller, J.S. Promoviendo la colaboración efectiva en MOOCs a través de aplicaciones móviles. In Proceedings of the 2015 CHILEAN Conference on Electrical, Electronics Engineering, Information and Communication Technologies, Santiago, Chile, 28–30 October 2015.
14. Goldie, J.G.S. Connectivism: A knowledge learning theory for the digital age? *Med. Teach.* **2016**, *38*, 1064–1069. [CrossRef] [PubMed]
15. Núñez-Martinez, M.; Borrás-Gené, O.; Fidalgo-Blanco, Á. Social community in MOOCs: Practical implications and outcomes. In Proceedings of the Second International Conference on Technological Ecosystems for Enhancing Multiculturality (TEEM), Salamanca, Spain, 1–3 October 2014; pp. 147–154.
16. García, M.I.; Díaz, C.G. Facebook como herramienta educativa en el contexto universitario. *Historia y Comunicación Social.* **2014**, *19*, 379–391.
17. Deterding, S.; Sicart, M.; Nacke, L.; O'Hara, K.; Dixon, D. Gamification. Using game-design elements in non-gaming contexts. In Proceedings of the CHI'11 Extended Abstracts on Human Factors in Computing Systems, Vancouver, BC, Canada, 7–12 May 2011; ACM: New York, NY, USA, 2011; pp. 2425–2428.
18. Werbach, K.; Hunter, D. *The Gamification Toolkit: Dynamics, Mechanics, and Components for the Win*; Wharton Digital Press: Philadelphia, PA, USA, 2015.
19. Kapp, K.M. *The Gamification of Learning and Instruction*; Wiley: San Francisco, CA, USA, 2013.
20. Vaibhav, A.; Gupta, P. Gamification of MOOCs for increasing user engagement. In Proceedings of the 2014 IEEE International Conference on MOOC, Innovation and Technology in Education (MITE), Patiala, India, 19–20 December 2014; pp. 290–295.
21. Borrás-Gené, O.; Martínez-Núñez, M.; Fidalgo-Blanco, Á. New challenges for the motivation and learning in engineering education using gamification in MOOC. *Int. J. Eng. Educ.* **2016**, *32*, 501–512.
22. Klemke, R.; Eradze, M.; Antonaci, A. The flipped MOOC: Using gamification and learning analytics in MOOC design—A conceptual approach. *Educ. Sci.* **2018**, *8*, 25. [CrossRef]
23. Huang, W.H.Y.; Soman, D. Gamification of Education. Available online: https://inside.rotman.utoronto.ca/behaviouraleconomicsinaction/files/2013/09/GuideGamificationEducationDec2013.pdf (accessed on 2 July 2019).
24. Kiryakova, G.; Angelova, N.; Yordanova, L. Gamification in education. In Proceedings of the 9th International Balkan Education and Science Conference, Edirne, Turkey, 16–18 October 2014.
25. Wang, L.; Hu, G.; Zhou, T. Semantic analysis of learners' emotional tendencies on online MOOC education. *Sustainability* **2018**, *10*, 1921. [CrossRef]
26. Hansch, A.; Newman, C.; Schildhauer, T. Fostering Engagement with Gamification: Review of Current Practices on Online Learning Platforms. Available online: https://papers.ssrn.com/sol3/papers.cfm?abstract_id=2694736 (accessed on 2 July 2019).
27. De-Marcos, L.; Domínguez, A.; Saenz-de-Navarrete, J.; Pagés, C. An empirical study comparing gamification and social networking on e-learning. *Comput. Educ.* **2014**, *75*, 82–91. [CrossRef]
28. Discover New Badges to Recognize Admins and Outstanding Members. Available online: https://www.facebook.com/community/education/blog/badges/ (accessed on 4 June 2019).
29. Telefónica Premia la Innovación Educativa de los MOOCs a Través de Míriadax. Available online: https://comunicacionmarketing.es/actualidad/24/01/2019/telefonica-premia-la-innovacion-educativa-de-los-moocs-a-traves-de-miriadax/5389.html (accessed on 4 June 2019).
30. ¡Conoce los Objetivos de Desarrollo Sostenible a Través del Nuevo MOOC de Miríadax! Available online: https://blogthinkbig.com/desarrollo-sostenible-miriadax (accessed on 27 June 2019).
31. García-Peñalvo, F.J.; Cruz-Benito, J.; Borrás-Gené, O.; Blanco, Á.F. Evolution of the Conversation and Knowledge Acquisition in Social Networks Related to a MOOC Course. In *Learning and Collaboration Technologies*; Zaphiris, P., Ioannou, A., Eds.; Springer: Berlin, German, 2015.

32. Pérez, A.B.; Borrás-Gené, O. Uso de Hangouts como recurso educativo en abierto en MOOC. In Proceedings of the IV Congreso Internacional sobre Aprendizaje, Innovación y Competitividad (CINAIC), Zaragoza, Spain, 4–6 October 2017.
33. Marsh, H.W. SEEQ: A Reliable, Valid, and Useful Instrument for Collecting Students' evaluations of University Teaching. *Br. J. Educ. Psychol.* **1982**, *52*, 77–95. [CrossRef]
34. Schroer, W. Defining, Managing, and Marketing to Generations, X, Y, and Z. *IAM* **2008**, *10*, 9.
35. Högberg, J.; Hamari, J.; Wästlund, E. Gameful Experience Questionnaire (GAMEFULQUEST): An instrument for measuring the perceived gamefulness of system use. *User Model. User-Adapt. Interac.* **2019**, *3*, 1–42. [CrossRef]
36. Dicheva, D.; Dichev, C.; Agre, G.; Angelova, G. Gamification in education: A systematic mapping study. *Educ. Technol. Soc.* **2015**, *18*, 75–88.

© 2019 by the authors. Licensee MDPI, Basel, Switzerland. This article is an open access article distributed under the terms and conditions of the Creative Commons Attribution (CC BY) license (http://creativecommons.org/licenses/by/4.0/).

Article

Video Games and Collaborative Learning in Education? A Scale for Measuring In-Service Teachers' Attitudes towards Collaborative Learning with Video Games

Marta Martín-del-Pozo *, Ana García-Valcárcel Muñoz-Repiso and Azucena Hernández Martín

Department of Didactics, Organization and Research Methods, University of Salamanca, Salamanca 37008, Spain
* Correspondence: mmdp@usal.es; Tel.: +34-923-294-500 (ext. 3401)

Received: 30 June 2019; Accepted: 1 August 2019; Published: 5 August 2019

Abstract: Students' motivation is a fundamental factor in the educational process, and can be facilitated through new methodologies and technologies, including gamification, video games, collaborative learning, or, in particular, the methodology called "collaborative learning with video games" (which is presented and can be understood as the implementation of educational activities in which students have to work together to achieve a goal, and the main resource of the activity is a video game). However, if teachers themselves are not motivated, or if they lack a positive attitude towards implementing these new methodologies, it will be difficult for students to feel motivated when approaching said resources. Therefore, it is important to know what teachers' attitudes towards them are. The aim of this research is the creation of an attitudes scale towards collaborative learning with video games, aimed at in-service primary school teachers. Different methodological steps were followed that made its construction possible, such as the analysis of items and the verification of their reliability, resulting in a rigorous attitudes scale of 33 items, with a reliability of $\alpha = 0.947$. This implies that the measurement instrument is validated and allows one to know the attitudes of in-service primary school teachers towards a new methodology related to the implementation of video games in education.

Keywords: video games; collaborative learning; education; teacher; attitudes; primary education; technology; ICT; Likert scale

1. Introduction

Students' motivation is a fundamental factor in the educational process at any educational stage. It can be facilitated through the implementation of new methodologies in the educational process, including project-based learning and problem-based learning, but also through the use of different technologies, such as robotics, augmented reality and virtual reality. In this text, we focus on gamification and video games, and in particular, on the methodology called "collaborative learning with video games", which deals with the use of video games in collaborative learning activities. These tools generate motivation and promote learning in students. However, if teachers do not have a positive attitude towards the implementation of these new tools, it will be difficult for students to feel motivated when these resources are implemented in their learning process. Therefore, it is important to know what teachers' attitudes towards them are [1–3].

The aim of this study was the creation of an attitudes scale directed towards collaborative learning with video games, aimed at in-service primary school teachers. With that in mind, this article is divided into several sections. We start with the theoretical framework, in which literature on the key concepts of the article is reviewed, including topics such as gamification and video games, motivation, learning gains and teachers' attitudes. A description of the study is presented in the following sections, including the methodology used and the results related to the creation of the instrument. Finally, the conclusions are addressed.

2. Theoretical Framework

Student motivation is a key factor for their learning, as indicated by various studies [4–6]. In fact, the implementation of gamification and video games in education has improved motivation for primary education students [7–9], secondary education students [10–12] and even higher education students [13–16]. In the same way, the implementation of collaborative learning methodologies has also contributed to students' motivation [6,17,18].

Furthermore, gamification and video games in education also enable learning gains for students in terms of knowledge, skills and attitudes, for different education system stages, including primary education [19–21], secondary education [22–24] and higher education [16,25,26]. At the same time, the implementation of collaborative learning methodologies also contributes to student learning [17,27–29].

This motivated us to create a methodology called "collaborative learning with video games", which brings together, in a single methodology, the advantages and criteria of implementing video games and collaborative learning in education. In that sense, it can be understood as the implementation of educational activities in which students have to work together, sharing responsibilities to achieve a goal (for instance, to do a task, to do a project, to complete a chart, to create a digital presentation, to write an essay, etc.), while discussing different perspectives and contributing with their ideas. The main resource of this activity is a video game [30]. It is important to highlight that collaborative learning between students can happen inside the video game, outside the game, or in both spaces (inside and outside the video game) depending on the type of educational strategy or activity the teacher chooses to implement. Educational experiences using this methodology are found in the literature [31–34]. Martín [35] provided further examples.

Considering these methodologies, students feel more motivated to face the educational process, and in turn, obtain learning gains. This poses a new question: are teachers motivated to implement these new technologies in their educational practices? What are teachers' attitudes, opinions and perspectives about it? If teachers are not motivated in their implementation, it will be difficult for students to feel motivated when these resources are presented as part of the learning process. In fact, as Tejedor and García-Valcárcel [3] said, one of the biggest factors that influences the integration of any pedagogical innovation, new methodology or new technological resource in educational practices, is the attitude of affected teachers. Therefore, it is important to understand teachers' viewpoints on these issues.

In this regard, we could find several studies referring to in-service and pre-service teachers' attitudes towards video games in education [1,2,36–38], and towards gamification in educational settings [39–42]. In general, research shows that teachers' attitudes towards these approaches are positive.

This article focuses on collaborative learning with video games; in particular, it reveals the creation and subsequent validation of a Likert-type attitude scale towards collaborative learning with video games, aimed at in-service primary education teachers.

3. Materials and Methods

The creation of an attitude scale requires a rigorous process in order to obtain an appropriate and validated instrument to measure the target attitude. At the same time, it is important to highlight that there are different types of attitude scales. In our specific case, we selected a Likert-type attitude scale. Likert-type attitude scales use a construction method that adapts to the measurement of different types of attitudes [43]. We took the ideas of multiple authors into account in creating this instrument [44,45].

Morales, Urosa and Blanco [44] synthesized the process of creation for this type of scale into specific stages, including the definition of the specific attitude to be measured as a key step. The next steps are: (1) the preparation of the instrument through the writing of several items and the preparation of additional information; (2) the obtainment of data from an adequate sample; (3) the item analysis, the calculation of reliability, the analysis of the scale content structure and the selection of definitive items.

It is fundamental to define the attitude to be measured, which, in our case, is teachers' attitudes towards collaborative learning with video games. This can be defined as the relatively stable predisposition of teachers to respond favorably or unfavorably to the implementation of educational activities in which students have to work together, sharing responsibilities to achieve a goal (for instance, to do a task, to do a project, to complete a chart, to create a digital presentation, or to write an essay), while discussing different perspectives and contributing with their ideas, where the main resource of this activity is a video game.

Regarding the writing of the items of the scale, it is necessary to create several sentences. As Morales [46] said, they are usually written in the form of opinions with which the person may or may not agree. In addition, other instruments that measure the same or similar attitudes can be taken into account for elaboration purposes. For that reason, we took the measurement instrument called "Semantic differential: learning through collaborative projects with Information and Communication Technologies (ICT)" [47] and the questionnaire "Opinion about collaborative learning methodology" [17] into account. A total of 75 preliminary items were prepared at this time. As Morales, Urosa and Blanco [44] pointed out, provisional items should be reviewed by more than one person, allowing for modification or elimination as needed. In our case, they were reviewed by an in-service primary education teacher (that is to say, a professional with a similar status to the final recipients of this instrument), by an expert in written comprehension and written composition, and an expert in educational technology. This process gave rise to 64 items. The following response mode was established: (1) strongly disagree (SD), (2) disagree (D), (3) indifferent (I), (4) agree (A), and (5) strongly agree (SA). As Morales, Urosa and Blanco [44] pointed out, additional information must be prepared, so other questions (specifically, 20 questions) were also incorporated into the complete questionnaire to provide more information including, for instance, age, gender, courses and disciplines taught this year, and university degrees finished.

Once we acquired a sufficient number of items and decided how to respond to them, it was necessary to validate our preliminary instrument by means of a methodologically appropriate procedure. Specifically, the instrument developed up to this point had to be subject to a content validation process by expert judgment. As Cabero and Barroso [48] said, the use of experts as a strategy to assess teaching materials, as well as data collection instruments (as in our case) or the methodologies used in the educational process, is quite common in the field of educational research. Six experts took part in our study. Considering that they were experts in more than one field, it is important to highlight that four were primary school teachers, five were experts in the implementation of information and communication technologies (ICT) or video games in educational settings, two were experts in collaborative learning, and two were experts in construction and validation of measuring instruments and research methods. Considering their own opinions and knowledge, the experts had to rate the validity of each item from 1 (very bad) to 5 (very good) in relation to the objective of the scale and the specific attitude to be measured. They could also submit other suggestions about the instrument in general and about specific items in a space created in the questionnaire for scale validation. Regarding the selection criteria of the items at this stage, that is, which preliminary items were kept in the instrument and which were not, we took into account the percentage of experts who considered that an item was good or very good (taking as criterion to be above 60% of experts), the average obtained (establishing as criterion an average equal to or greater than 4), and the specific comments and suggestions provided by the experts. In addition, in the case that some of the items did not meet these criteria, we took the utility of that item on the scale into account, and were able to keep those items for the next step in which they were used with a sample. As a result of this process, we kept 57 items of the initial 64, namely those that applied to in-service primary school teachers.

We then went to the next step of creating the scale, which involved obtaining data from appropriate samples. Specifically, the sample included 223 Spanish in-service primary school teachers. With this data, we proceeded to carry out item analysis, reliability calculation, analysis of the scale content structure and final selection of the items using the statistical software SPSS 22 (IBM Corporation, Armonk, NY, USA). These aspects will be presented in the following section.

4. Data Analysis and Results

Firstly, we present the results that show the sample characterization. The total sample was 223 Spanish in-service primary school teachers, and 113 were men (50.7%) and 110 were women (49.3%). The respondents were 21–60 years old, the statistical mode was 38 years old (5.4% of the sample) and the mean was 36.09. In regards to what levels they teach (considering that primary education in Spain consists of six years, from 6 to 12 years old), 79 teachers taught the first year, 71 the second year, 78 the third year, 81 the fourth year, 87 the fifth year, and 87 the sixth year. It is important to highlight that in Spain the same teacher can work across different age levels depending on the discipline. In regards to their university-level education, excluding their undergraduate degree to be a primary school teacher, 17 teachers (7.6%) were enrolled in a master's degree and 36 teachers (16.1%) held a master's degree at the time of answering the questionnaire. Furthermore, nine teachers (4%) were enrolled in a PhD, and five teachers (2.24%) had finished their doctoral studies.

Once we obtained data from a sample, the next step was to analyze the items and the verification of reliability. We had to check whether each item from the initial version measured the same attitude as the other items. This is fundamental, in order to know if it is possible to sum its specific item score in a total score that supposedly measures the attitude that we want to study, taking into account that the total score of each person is one that will later be interpreted. This check was done through item analysis, and we used the item–total correlation procedure, which is, properly speaking, the correlation of each item with the sum of all others, or the correlation of each item with the total minus the item, i.e., the corrected item–total correlation [44]. We wanted to check whether scoring high on an item means, in fact, getting a high total score on the rest of the scale. When selecting our items, we had to take into account the fact that those with scores that correlated most highly with the sum of all the others are those that have more in common, and we can assume they measure the same as the rest. However, the items that show non-significant or very low correlations in relation to the rest of the items must be eliminated from the scale [44]. In addition, it must be taken into account that the process does not have to be automatic, rather the researchers' ideas about what they are trying to measure need to be considered, so conceptual criteria must also be taken into account. Thus, in Appendix A we show the total-element statistics of the items in the 57 item scale version, and in Appendix B we show the total-element statistics of the items in the final version of 33 items; that is, these are the definitive 33 items that will be part of the scale (the complete items can be seen in Appendix C, translated from Spanish). It is important to highlight that definitive items are numbered in the appendices and in tables according to their sequence in the final instrument and in Appendix C (from 'item 1' to 'item 33'), while the eliminated items have been named through letters (from 'eliminated item A' to 'eliminated item X').

Considering the above, this gives rise to an attitude scale of 33 items, with a reliability of $\alpha = 0.947$. Furthermore, we carried out factorial analysis as a method to check the construct validity of the final version with the 33 items. As Morales, Urosa and Blanco [44] pointed out, factorial analysis with the rotated factors allows us to appreciate whether we are measuring what we say we measure, by clarifying the matters that underlie several variables, what items are defining each factor and how these factors relate to each other, helping us to clarify the structure of the instrument and the construct. In Table 1, we show data about the factors extracted in the analysis, and in Table 2, we show the rotated component matrix for the 33 item final version of the scale.

Table 1. Data about the factors extracted in the analysis (in the final version with 33 items).

Component	Total Variance Explained [1]								
	Initial Eigenvalues			Extration Sums of Squared Loadings			Rotation Sums of Squared Loadings		
	Total	% of Variance	Cumulative %	Total	% of Variance	Cumulative %	Total	% of Variance	Cumulative %
1	12.837	38.900	38.900	12.837	38.900	38.900	6.402	19.400	19.400
2	2.115	6.410	45.310	2.115	6.410	45.310	4.360	13.211	32.612
3	1.866	5.656	50.966	1.866	5.656	50.966	2.742	8.308	40.920
4	1.129	3.421	54.386	1.129	3.421	54.386	2.415	7.319	48.239
5	1.100	3.335	57.721	1.100	3.335	57.721	2.106	6.383	54.621
6	1.043	3.161	60.882	1.043	3.161	60.882	2.066	6.261	60.882
7	0.963	2.919	63.802						
8	0.940	2.849	66.651						
9	0.783	2.373	69.024						
10	0.754	2.284	71.308						
11	0.710	2.150	73.458						
12	0.689	2.088	75.546						
13	0.639	1.937	77.483						
14	0.613	1.859	79.342						
15	0.577	1.749	81.091						
16	0.546	1.656	82.746						
17	0.501	1.518	84.265						
18	0.466	1.412	85.677						
19	0.457	1.385	87.062						
20	0.442	1.340	88.402						
21	0.416	1.262	89.664						
22	0.399	1.210	90.874						
23	0.390	1.180	92.054						
24	0.348	1.053	93.107						
25	0.329	0.997	94.104						
26	0.307	0.929	95.033						
27	0.282	0.853	95.886						
28	0.278	0.842	96.728						
29	0.258	0.782	97.509						
30	0.235	0.711	98.221						
31	0.229	0.694	98.915						
32	0.211	0.640	99.555						
33	0.147	0.445	100.000						

[1] Extraction method: principal component analysis.

Table 2. Rotated component matrix for the final version of the instrument with 33 items.

Rotated Component Matrix [2]						
	Component					
	1	2	3	4	5	6
Item 4	0.730					
Item 22	0.702					
Item 12	0.690					0.305
Item 18	0.685	0.311				
Item 28	0.680					
Item 6	0.663					
Item 10	0.611				0.309	
Item 19	0.604				0.333	
Item 17	0.593					
Item 20	0.545					
Item 14	0.535					0.440
Item 24	0.529	0.369				
Item 23	0.528	0.385				
Item 33		0.727				
Item 25		0.722				
Item 21	0.355	0.711				
Item 27	0.491	0.616				
Item 2		0.571				
Item 31		0.551				

Table 2. Cont.

	Rotated Component Matrix [2]					
	Component					
	1	2	3	4	5	6
Item 32	0.313	0.501	0.316		0.357	
Item 1			0.690			
Item 29			0.640			
Item 30		0.320	0.616			0.311
Item 26		0.391	0.508			
Item 13				0.866		
Item 5	0.314			0.784		
Item 8			0.416	0.610		
Item 9					0.704	
Item 16					0.633	
Item 11	0.443				0.463	
Item 15						0.664
Item 3				0.348		0.573
Item 7			0.349			0.526

[2] Extraction method: principal component analysis. Rotation method: varimax with Kaiser normalization.

As we can see in Table 1, six factors were extracted that explain 60.882% of the total variance, taking into account that, to determine the number of factors that have to be extracted, those components with eigenvalues greater than 1 are conserved [49]. This also fulfils what was indicated by Nunnally [43], because it is necessary to eliminate the factors in which no variable has a weight superior to 0.30, and, as can be appreciated in Table 2, all factors have some variable with a weight greater than this. In addition, it is necessary to take into account only those factors that are defined by at least three items [44], which is what happens with our six factors (as can be seen in Table 2 in the different columns for each component).

In order to interpret factor structure, we examined the saturations that, in each factor, obtained the items of the scale [49] according to the results in Table 2. We attended mainly to those items with the largest weights [44] and chose, in those cases where there are items that were saturated in more than one factor, to place them with the factor in which they saturated the most.

As we can see in Table 2, the first factor is integrated by items 4, 6, 10, 12, 14, 17, 18, 19, 20, 22, 23, 24 and 28, and explains 38.9% of the variance (as we can see in the third column of Table 1). We denominated it "educational possibilities", because the items highlight the educational possibilities of collaborative learning with video games. For instance, the greater the interaction between the teacher and the students, the greater the students' autonomy in their learning, the development of students' capacity for initiative and the possibility to explore ideas and concepts more fully.

The second factor is integrated by items 2, 21, 25, 27, 31, 32 and 33, and explains 6.41% of the variance. We called it "positive disposition to implement activities" by incorporating those items that include formulations showing interest, inclination or attraction towards the approach of collaborative learning activities with video games, for example, showing interest to collaborate with other teachers who implement these kinds of activities or showing interest to work in a school where this methodology was supported.

The third factor is integrated by items 1, 26, 29 and 30, and explains 5.656% of the variance. We called it "denial as educational methodology" because the items are related to the rejection of collaborative learning with video games as a possible methodology to be applied in educational practices, indicating that implementing this methodology is impossible and inappropriate. Taking this into account, it is important to highlight that all the items in this factor are negative and, in the analysis, it was necessary to reverse the score.

The fourth factor is integrated by items 5, 8 and 13, and explains 3.421% of the variance. We denominated it "concerns about neglecting the learning" by incorporating those items that are related to teachers' concerns about the implementation of this kind of methodology and the problem of neglecting or not giving the required importance to learning by the students, such as taking learning lightly and not putting effort into educational tasks. As for the previous factor, all the items in this factor are also negative and, in the analysis, it was necessary to reverse the score.

The fifth factor is integrated by items 9, 11 and 16, and explains 3.335% of the variance. We denominated it "useful and inclusive learning strategy" by incorporating those formulations related to the idea of collaborative learning with video games methodology as a learning strategy that allows the inclusion of all students and that allows learning relevant matters for their lives in the complex and diverse world in which we live.

Finally, the sixth factor is integrated by items 3, 7 and 15, and explains 3.161% of the variance. We called it "teacher denial due to loss of time" by incorporating those formulations in which the teacher rejects this approach, considering it a waste of time in terms of class time and personal time. In this case, the three items are negative and, in the analysis, it was also necessary to reverse the score.

The analysis of items shown, and the factorial analysis carried out, led us to confirm the selection of the 33 items as elements for the final version of the scale. The scale has a reliability of $\alpha = 0.947$, with the reliability of each factor as following:

- Factor 1 "educational possibilities": $\alpha = 0.921$.
- Factor 2 "positive disposition to implement activities": $\alpha = 0.876$
- Factor 3 "denial as educational methodology": $\alpha = 0.762$
- Factor 4 "concerns about neglecting the learning": $\alpha = 0.814$
- Factor 5 "useful and inclusive learning strategy": $\alpha = 0.662$
- Factor 6 "teacher denial due to loss of time": $\alpha = 0.696$.

Finally, it should also be noted that, although we tried to have the same number of items in the affective, cognitive and behavioural fields, the scale has the following final structure:

- Twelve items related to the affective field (items 2, 3, 5, 7, 8, 13, 21, 25, 26, 27, 29 and 33): $\alpha = 0.873$
- Thirteen items related to the cognitive field (items 1, 4, 6, 9, 10, 12, 14, 17, 18, 20, 22, 24 and 28): $\alpha = 0.904$
- Eight items related to the behavioural field (items 11, 15, 16, 19, 23, 30, 31 and 32): $\alpha = 0.832$.

5. Conclusions

Video games, gamification and collaborative learning are elements that can be implemented in education because they generate student motivation and contribute to student learning in different education system stages, as we saw in the theoretical framework of this article. However, the generation of students' motivation and students' learning when teachers implement these resources in education can also be influenced by the teachers' attitudes towards these elements. In fact, the teacher's attitude towards a new resource or methodology is one of the main factors that contributes to its implementation in educational practices. Therefore, it becomes relevant to know teachers' attitudes towards new methodologies and resources, which in our case is collaborative learning with video games methodology, which brings together in a single methodology the advantages and criteria of implementing video games and collaborative learning in education. For that reason, in order to know and analyze these attitudes, validated and reliable instruments are required, which must be built with a rigorous construction process. In this text, we showed the creation of an attitudes scale towards collaborative learning with video games aimed at in-service primary school teachers, developed in a rigorous way that enables us to know teachers' attitudes towards this specific methodology. We believe that the availability of this instrument in the scientific community will contribute to the study of this variable for other researchers who are interested in the area, as well as the development of other measurement instruments related to the field of video games, gamification and collaborative learning.

Author Contributions: Conceptualization, M.M.-d.-P.; methodology, M.M.-d.-P, A.G.-V.M.-R. and A.H.M.; formal analysis, M.M.-d.-P. and A.G.-V.M.-R.; investigation, M.M.-d.-P., A.G.-V.M.-R. and A.H.M.; writing—original draft preparation, M.M.-d.-P.; writing—review and editing, M.M.-d.-P., A.G.-V.M.-R. and A.H.M.; funding acquisition, M.M.-d.-P.

Funding: In terms of the first author, this research was made possible through the funding of a FPU predoctoral grant (FPU13/02194) from the Ministry of Education, Culture and Sport of Spain.

Acknowledgments: We would like to thank the experts who participated in the revision of the instrument and the in-service primary school teachers who answered the initial version of the attitudes scale.

Conflicts of Interest: The authors declare no conflict of interest. The funders had no role in the design of the study; in the collection, analyses, or interpretation of data; in the writing of the manuscript, or in the decision to publish the results.

Appendix A

Table A1. Total-element statistics of the items in the 57 item scale version.

Items [3]	Scale Mean If Item Deleted	Scale Variance If Item Deleted	Corrected Item-Total Correlation	Cronbach's Alpha If Item Deleted
Eliminated item A	220.63	662.559	0.602	0.958
Item 2	220.62	665.282	0.551	0.958
Eliminated item B	220.87	668.153	0.483	0.958
Eliminated item C	220.62	672.471	0.386	0.959
Eliminated item D	221.07	664.586	0.448	0.958
Item 3	220.56	664.834	0.664	0.958
Eliminated item E	220.41	671.550	0.437	0.958
Eliminated item F	220.96	661.998	0.529	0.958
Eliminated item G	220.65	669.445	0.490	0.958
Item 16	220.64	667.935	0.582	0.958
Item 9	220.61	669.943	0.442	0.958
Eliminated item H	220.93	665.716	0.456	0.958
Eliminated item I	221.23	665.819	0.592	0.958
Eliminated item J	221.12	669.692	0.423	0.958
Eliminated item K	221.36	665.303	0.518	0.958
Item 14	221.05	664.462	0.577	0.958
Item 7	220.55	665.627	0.561	0.958
Eliminated item L	220.50	671.828	0.507	0.958
Eliminated item M	220.79	667.020	0.475	0.958
Item 17	221.23	663.963	0.611	0.958
Item 15	220.53	672.845	0.501	0.958
Item 12	221.16	662.172	0.575	0.958
Eliminated item N	220.65	672.014	0.433	0.958
Item 11	220.79	669.255	0.524	0.958
Item 10	221.18	666.826	0.479	0.958
Item 13	221.61	660.157	0.485	0.958
Item 19	220.91	667.343	0.554	0.958
Item 5	221.61	654.418	0.575	0.958
Item 23	220.79	666.146	0.593	0.958
Item 20	220.95	665.993	0.575	0.958
Item 18	220.99	664.225	0.629	0.958
Item 22	221.14	661.394	0.615	0.958
Item 27	220.74	660.407	0.705	0.957
Item 21	220.69	662.710	0.671	0.958
Item 6	220.93	664.099	0.639	0.958
Item 32	220.70	670.357	0.546	0.958
Eliminated item O	221.51	682.638	0.082	0.961

Table A1. Cont.

Items [3]	Scale Mean If Item Deleted	Scale Variance If Item Deleted	Corrected Item-Total Correlation	Cronbach's Alpha If Item Deleted
Item 24	220.86	663.345	0.657	0.958
Item 4	221.13	662.594	0.566	0.958
Item 28	220.95	661.240	0.669	0.958
Eliminated item P	221.62	662.156	0.461	0.958
Eliminated item Q	221.37	663.378	0.541	0.958
Eliminated item R	221.19	660.165	0.615	0.958
Item 8	221.37	652.107	0.646	0.958
Eliminated item S	221.32	666.831	0.421	0.959
Eliminated item T	221.23	666.808	0.447	0.958
Eliminated item U	220.68	671.652	0.439	0.958
Item 26	220.88	662.683	0.560	0.958
Item 25	220.75	662.943	0.581	0.958
Item 29	220.95	662.006	0.568	0.958
Eliminated item V	220.70	664.472	0.593	0.958
Item 30	220.71	662.721	0.604	0.958
Item 31	220.88	663.152	0.630	0.958
Eliminated item W	220.69	669.108	0.488	0.958
Item 1	220.85	664.544	0.511	0.958
Item 33	220.68	666.913	0.565	0.958
Eliminated item X	220.65	663.824	0.585	0.958

[3] Cronbach's Alpha with 57 items: 0.959.

Appendix B

Table A2. Total-element statistics of the items in the final version of 33 items.

Items [4]	Scale Mean If Item Deleted	Scale Variance If Item Deleted	Corrected Item-Total Correlation	Cronbach's Alpha If Item Deleted
Item 1	126.69	256.478	0.475	0.947
Item 2	126.46	256.384	0.536	0.946
Item 3	126.39	257.132	0.602	0.946
Item 4	126.97	253.008	0.614	0.945
Item 5	127.45	250.159	0.545	0.946
Item 6	126.77	254.438	0.677	0.945
Item 7	126.39	256.724	0.541	0.946
Item 8	127.21	249.408	0.596	0.946
Item 9	126.44	258.554	0.454	0.947
Item 10	127.02	255.779	0.521	0.946
Item 11	126.63	257.846	0.553	0.946
Item 12	127.00	253.189	0.606	0.945
Item 13	127.45	254.627	0.427	0.948
Item 14	126.89	255.253	0.587	0.946
Item 15	126.37	261.685	0.453	0.947
Item 16	126.48	258.088	0.564	0.946
Item 17	127.07	254.887	0.624	0.945
Item 18	126.83	254.574	0.664	0.945
Item 19	126.75	256.054	0.608	0.945
Item 20	126.79	255.798	0.604	0.945

Table A2. *Cont.*

Items [4]	Scale Mean If Item Deleted	Scale Variance If Item Deleted	Corrected Item-Total Correlation	Cronbach's Alpha If Item Deleted
Item 21	126.53	253.989	0.692	0.945
Item 22	126.98	252.099	0.672	0.945
Item 23	126.63	256.179	0.611	0.945
Item 24	126.70	254.671	0.665	0.945
Item 25	126.59	254.450	0.585	0.946
Item 26	126.72	256.447	0.484	0.947
Item 27	126.58	251.875	0.753	0.944
Item 28	126.78	252.557	0.710	0.945
Item 29	126.78	254.197	0.559	0.946
Item 30	126.55	254.483	0.602	0.945
Item 31	126.72	254.600	0.635	0.945
Item 32	126.53	258.484	0.581	0.946
Item 33	126.52	257.323	0.553	0.946

[4] Cronbach's Alpha with 33 items: 0.947.

Appendix C

Items of "collaborative learning with video games attitudes scale" for in-service primary school teachers (translated from Spanish).

1. Implementing video games for collaborative learning in educational practices is impossible.
2. I would like to implement collaborative learning activities with video games in educational practices.
3. If I implemented collaborative learning activities with video games in educational practices, I would feel that I am wasting class time.
4. Collaborative learning with video games allows for greater interaction between the teacher and his/her students.
5. I worry that collaborative learning activities with video games encourage students not to put effort into educational tasks and activities.
6. Collaborative learning with video games allows students to jointly build knowledge about curricular content.
7. I think receiving training in collaborative learning with video games is a waste of time.
8. I worry that collaborative learning with video games is a distraction from the course syllabus.
9. Collaborative learning with video games is a good strategy for the inclusion of students with special education needs.
10. When working with video games in groups, students would pay attention to the opinions of other students.
11. I would implement collaborative learning activities with video games to help students learn to share responsibilities.
12. By working collaboratively with video games in educational practices, students would relate to each other more easily.
13. I worry that collaborative learning with video games encourages students to take learning lightly.
14. Collaborative learning with video games allows students to learn to work autonomously.
15. If my students asked me to carry out collaborative learning activities with video games in educational practices, I would refuse.
16. I would implement collaborative learning activities with video games to help students develop useful life skills.
17. Collaborative learning activities with video games help to explore ideas and concepts more fully.

18. Students have greater autonomy in their learning when they take part in collaborative learning activities with video games.
19. I would implement collaborative learning activities with video games to increase the students' self-esteem.
20. Students would put more effort to share knowledge among them if they worked collaboratively with video games.
21. I would like to encourage the curiosity of students through collaborative learning with video games.
22. Video games facilitate the implementation of collaborative activities with students.
23. I would implement collaborative learning activities with video games to develop the students' capacity for initiative.
24. When working collaboratively with video games, the explanations given among the members of the group facilitate the understanding of the concepts.
25. I would like to work in a school where the implementation of collaborative learning activities with video games with students was supported.
26. I would be overwhelmed if I had to implement collaborative learning activities with video games with students.
27. I would like to develop the students' creativity through collaborative learning with video games.
28. When working collaboratively with video games in educational practices, the interaction generated with classmates increases the level of student learning.
29. I do not believe that collaborative learning with video games is an appropriate classroom methodology that improves education.
30. If I had to implement new activities in the educational practices, they would never be collaborative learning activities with video games.
31. If there were sufficient resources within the school, I would frequently implement collaborative learning activities with video games.
32. I would implement collaborative learning activities with video games to facilitate the students to learn the course syllabus.
33. I would like to collaborate with other teachers who implement collaborative learning activities with video games in their educational practices.

References

1. Noraddin, E.M.; Kian, N.T. Academics' attitudes toward using digital games for learning & teaching in Malaysia. *Malays. Online J. Educ. Technol.* **2014**, *2*, 1–21.
2. Pastore, R.S.; Falvo, D.A. Video games in the classroom: Pre- and in-service teachers' perceptions of games in the K-12 classroom. *Int. J. Instr. Technol. Distance Learn.* **2010**, *7*, 49–57.
3. Tejedor, F.J.; García-Valcárcel, A. Competencias de los profesores para el uso de las TIC en la enseñanza. Análisis de sus conocimientos y actitudes. *Rev. Esp. Pedagog.* **2006**, *64*, 21–43.
4. Anjomshoa, L.; Sadighi, F. The Importance of Motivation in Second Language Acquisition. *IJSELL* **2015**, *3*, 126–137.
5. Feng, R.; Chen, H. An Analysis on the Importance of Motivation and Strategy in Postgraduates English Acquisition. *Engl. Lang. Teach.* **2009**, *2*, 93–97. [CrossRef]
6. Serrano-Cámara, L.M.; Paredes-Velasco, M.; Alcover, C.M.; Velazquez-Iturbide, J.A. An evaluation of students' motivation in computer-supported collaborative learning of programming concepts. *Comput. Hum. Behav.* **2014**, *31*, 499–508. [CrossRef]
7. Gooch, D.; Vasalou, A.; Benton, L.; Khaled, R. Using Gamification to Motivate Students with Dyslexia. In Proceedings of the 2016 CHI Conference on Human Factors in Computing Systems, San Jose, CA, USA, 7–12 May 2016; pp. 969–980.
8. Hill, V. Digital citizenship through game design in Minecraft. *New Libr. World* **2015**, *116*, 369–382. [CrossRef]
9. Ordiz, T. Gamificación: La vuelta al mundo en 80 días. *Infanc. Educ. Aprendiz.* **2017**, *3*, 397–403. [CrossRef]

10. Morillas, C.; Muñoz-Organero, M.; Sánchez, J. Can Gamification Improve the Benefits of Student Response Systems in Learning? An Experimental Study. *IEEE Trans. Emerg. Top. Comput.* **2016**, *4*, 429–438. [CrossRef]
11. Quintero, L.E.; Jiménez, F.; Area, M. Más allá del libro de texto. La gamificación mediada con TIC como alternativa de innovación en Educación Física. *Retos Nuevas Tend. Educ. física deporte recreación* **2018**, *34*, 343–348.
12. Hong, G.Y.; Masood, M. Effects of Gamification on Lower Secondary School Students' Motivation and Engagement. *Int. J. Educ. Pedagog. Sci.* **2014**, *8*, 3765–3772.
13. Corchuelo-Rodríguez, C.A. Gamificación en Educación Superior: Experiencia innovadora para motivar estudiantes y dinamizar contenidos en el aula. *EDUTEC* **2018**, *63*, 29–41. [CrossRef]
14. Çağlar, S.; Kocadere, S.A. Possibility of Motivating Different Type of Players in Gamified Learning Environments. In Proceedings of the EDULEARN16 Conference, Barcelona, Spain, 4–6 July 2016; pp. 1987–1994.
15. Landers, R.N.; Callan, R.C. Casual Social Games as Serious Games: The Psychology of Gamification in Undergraduate Education and Employee Training. In *Serious Games and Edutainment Applications*; Ma, M., Oikonomou, A., Jain, L., Eds.; Springer: London, UK, 2011; pp. 399–423.
16. Topîrceanu, A. Gamified learning: A role-playing approach to increase student in-class motivation. *Procedia Comput. Sci.* **2017**, *112*, 41–50. [CrossRef]
17. García-Valcárcel, A.; Hernández, A.; Recamán, A. La metodología del aprendizaje colaborativo a través de las TIC: Una aproximación a las opiniones de profesores y alumnos. *Rev. Complut. Educ.* **2012**, *23*, 161–188. [CrossRef]
18. Jones, B.D.; Epler, C.M.; Mokri, P.; Bryant, L.H.; Paretti, M.C. The Effects of a Collaborative Problem-based Learning Experience on Students' Motivation in Engineering Capstone Courses. *Interdiscip. J. Probl.-Based Learn.* **2013**, *7*, 34–71. [CrossRef]
19. Capell, N.; Tejada, J.; Bosco, A. Los videojuegos como medio de aprendizaje: Un estudio de caso en matemáticas en Educación Primaria. *Píxel-Bit* **2017**, *51*, 133–150. [CrossRef]
20. Chen, H.R.; Jian, C.H.; Lin, W.S.; Yang, P.C.; Chang, H.Y. Design of Digital Game-based Learning in Elementary School Mathematics. In Proceedings of the 2014 7th International Conference on Ubi-Media Computing and Workshops, Ulaanbaatar, Mongolia, 12–14 July 2014; pp. 322–325.
21. Miller, D.J.; Robertson, D.P. Educational benefits of using game consoles in a primary classroom: A randomised controlled trial. *Br. J. Educ. Technol.* **2011**, *42*, 850–864. [CrossRef]
22. Brom, C.; Preuss, M.; Klement, D. Are educational computer micro–games engaging and effective for knowledge acquisition at high–schools? A quasi–experimental study. *Comput. Educ.* **2011**, *57*, 1971–1988. [CrossRef]
23. Huizenga, J.; Admiraal, W.; Akkerman, S.; Ten Dam, G. Mobile game-based learning in secondary education: Engagement, motivation and learning in a mobile city game. *J. Comput. Assist. Learn.* **2009**, *25*, 332–344. [CrossRef]
24. Cheng, M.-T.; Lin, Y.-W.; She, H.-C. Learning through playing Virtual Age: Exploring the interactions among student concept learning, gaming performance, in-game behaviors, and the use of in-game characters. *Comput. Educ.* **2015**, *86*, 18–29. [CrossRef]
25. Perini, S.; Luglietti, R.; Margoudi, M.; Oliveira, M.; Taisch, M. Learning and motivational effects of digital game-based learning (DGBL) for manufacturing education—The Life Cycle Assessment (LCA) game. *Comput. Ind.* **2018**, *102*, 40–49. [CrossRef]
26. Tan, A.J.Q.; Lee, C.C.S.; Lin, P.Y.; Cooper, S.; Lau, L.S.T.; Chua, W.L.; Liaw, S.Y. Designing and evaluating the effectiveness of a serious game for safe administration of blood transfusion: A randomized controlled trial. *Nurse Educ. Today* **2017**, *55*, 38–44. [CrossRef]
27. Awedh, M.; Mueen, A.; Zafar, B.; Manzoor, U. Using Socrative and Smartphones for the support of collaborative learning. *IJITE* **2014**, *3*, 17–24. [CrossRef]
28. Casillas, S.; Martín, J.; Martín, M.; Hernández, M.J. Proyecto "Empléate". In *Proyectos de Trabajo Colaborativo con TIC*; García-Valcárcel, A., Ed.; Síntesis: Madrid, Spain, 2015; pp. 219–230.
29. Terenzini, P.T.; Cabrera, A.F.; Colbeck, C.L.; Parente, J.M.; Bjorklund, S.A. Collaborative Learning vs. Lecture/Discussion: Students' Reported Learning Gains. *J. Eng. Educ.* **2001**, *90*, 123–130. [CrossRef]

30. Martín, M.; García-Valcárcel, A.; Hernández, A. Video games in teacher training: Design, implementation and assessment of an educational proposal. In Proceedings of the Fourth International Conference on Technological Ecosystems for Enhancing Multiculturality (TEEM '16), Salamanca, Spain, 2–4 November 2016; pp. 1147–1154.
31. Henderson, L.; Klemes, J.; Eshet, Y. Just Playing a Game? Educational Simulation Software and Cognitive Outcomes. *J. Educ. Comput. Res.* **2000**, *22*, 105–129. [CrossRef]
32. Lester, J.C.; Spires, H.A.; Nietfeld, J.L.; Minogue, J.; Mott, B.W.; Lobene, E.V. Designing game-based learning environments for elementary science education: A narrative-centered learning perspective. *Inf. Sci.* **2014**, *264*, 4–18. [CrossRef]
33. Pareto, L.; Haake, M.; Lindström, P.; Sjödén, B.; Gulz, A. A teachable-agent based game affording collaboration and competition: Evaluating math comprehension and motivation. *Educ. Technol. Res.* **2012**, *60*, 723–751. [CrossRef]
34. Sung, H.Y.; Hwang, G.J. A collaborative game-based learning approach to improving students' learning performance in science courses. *Comput. Educ.* **2013**, *63*, 43–51. [CrossRef]
35. Martín, M. Videojuegos y aprendizaje colaborativo. Experiencias en torno a la etapa de Educación Primaria. *EKS* **2015**, *16*, 69–89.
36. Jenny, S.E.; Hushman, G.F.; Hushman, C.J. Pre-service teachers' perceptions of motion-based video gaming in physical education. *Int. J. Technol. Teach. Learn.* **2013**, *9*, 96–111.
37. Martín, M.; Basilotta, V.; García-Valcárcel, A. An approach to Spanish Primary School Teachers' attitudes towards collaborative learning with video games and the influence of teacher training. In Proceedings of the Fourth International Conference on Technological Ecosystems for Enhancing Multiculturality (TEEM '16), Salamanca, Spain, 2–4 November 2016; pp. 715–719.
38. Proctor, M.D.; Marks, Y. A survey of exemplar teachers' perceptions, use, and access of computer-based games and technology for classroom instruction. *Comput. Educ.* **2013**, *62*, 171–180. [CrossRef]
39. Alabbasi, D. Exploring Teachers Perspectives towards Using Gamification Techniques in Online Learning. *Turk. Online J. Educ. Tojet* **2018**, *17*, 34–45.
40. Hung, A.C.Y.; Zarco, E.; Yang, M.; Dembicki, D.; Kase, M. Gamification in the wild: Faculty perspectives on gamifying learning in higher education. *Issues Trends Educ. Technol.* **2017**, *5*, 4–22.
41. Martí-Parreño, J.; Seguí-Mas, D.; Seguí-Mas, E. Teachers' Attitude towards and Actual Use of Gamification. *Procedia Soc. Behv.* **2016**, *228*, 682–688. [CrossRef]
42. Sánchez-Mena, A.; Martí-Parreño, J. Drivers and Barriers to Adopting Gamification: Teachers' Perspectives. *Electron. J. e-Learn.* **2017**, *15*, 434–443.
43. Nunnally, J.C. *Psychometric Theory*; McGraw-Hill: New York, NY, USA, 1978.
44. Morales, P.; Urosa, B.; Blanco, A. *Construcción de Escalas de Actitudes "Tipo Likert": Una Guía Práctica*; La Muralla y Hespérides: Madrid and Salamanca, Spain, 2003.
45. Tejedor, F.J.; García-Valcárcel, A.; Prada, S. Medida de actitudes del profesorado universitario hacia la integración de las TIC. *Comunicar* **2009**, *33*, 115–124.
46. Morales, P. *Medición de Actitudes en Psicología y Educación: Construcción de Escalas y Problemas Metodológicos*; Publicaciones de la Universidad Pontificia Comillas: Madrid, Spain, 2006.
47. García-Valcárcel, A. *Proyectos de Trabajo Colaborativo con TIC*; Síntesis: Madrid, Spain, 2015.
48. Cabero, J.; Barroso, J. La utilización del juicio de experto para la evaluación de TIC: El coeficiente de competencia experta. *Bordón* **2013**, *65*, 25–38. [CrossRef]
49. García, E.; Gil, J.; Rodríguez, G. *Análisis Factorial*. *Cuadernos de Estadística, 7*; Editorial La Muralla: Madrid, Spain, 2000.

© 2019 by the authors. Licensee MDPI, Basel, Switzerland. This article is an open access article distributed under the terms and conditions of the Creative Commons Attribution (CC BY) license (http://creativecommons.org/licenses/by/4.0/).

Article

The Effects of Gamification in Online Learning Environments: A Systematic Literature Review

Alessandra Antonaci *,†, Roland Klemke and Marcus Specht

Welten Institute–Research Centre for Learning, Teaching and Technology, Open University of The Netherlands, 6401 DL Heerlen, The Netherlands
* Correspondence: alessandra.antonaci@ou.nl
† Current address: Valkenburgerweg, 177, 6401 DL Heerlen, The Netherlands.

Received: 10 June 2019; Accepted: 8 August 2019; Published: 12 August 2019

Abstract: Gamification has recently been presented as a successful strategy to engage users, with potential for online education. However, while the number of publications on gamification has been increasing in recent years, a classification of its empirical effects is still missing. We present a systematic literature review conducted with the purpose of closing this gap by clarifying what effects gamification generates on users' behaviour in online learning. Based on the studies analysed, the game elements most used in the literature are identified and mapped with the effects they produced on learners. Furthermore, we cluster these empirical effects of gamification into six areas: performance, motivation, engagement, attitude towards gamification, collaboration, and social awareness. The findings of our systematic literature review point out that gamification and its application in online learning and in particular in Massive Online Open Courses (MOOCs) are still a young field, lacking in empirical experiments and evidence with a tendency of using gamification mainly as external rewards. Based on these results, important considerations for the gamification design of MOOCs are drawn.

Keywords: gamification; game elements; online learning; MOOCs; empirical studies; systematic literature review

---

1. Introduction

It is often stated that gamification is a winning approach to 'motivate' people. Despite this belief, and broad field of application, its empirical validation is still to be proven. Furthermore, the field lacks an understanding of the main effects of gamification on short-, middle- and long-term. Gamification design, implementation and effectiveness are related to the audience and the context of application; nevertheless, a generalisation and analysis of current results and studies is needed to further improve the field. In order to be effective, designers of gamification need to be aware of what the outcome/s of a specific game element could be in a certain scenario and audience. This systematic literature review is a contribution to the field by providing an overview and map of the game elements most used in online learning environments, including Massive Open Online Courses (MOOCs), and their empirically proven effect/s on human behaviours.

The focus on online learning and MOOCs is due to the fact that online learning in general has several points in common and a number of differences with MOOCs in particular. To some extent, the latter can be seen as a subcategory of the first. However, MOOCs are different from online learning because they can help to scale up open learning opportunities [1]. MOOCs, taking advantage of cutting-edge technologies and the Internet, are reaching a massive audience and, until the recent past, they were free [2]. The MOOC phenomenon exploded in 2012, which was declared the year of MOOCs [3]. In almost the same period, in 2011, the gamification phenomenon reached its first peak, with the elaboration of the most known definition elaborated by Deterding et al. [4] ('gamification' consists of the transfer of game design elements to a non-game scenario). Since then, gamification has been applied to different fields (see Section 2); however, probably due to the concurrent development of both MOOCs and gamification, there are few empirical applications of gamification to MOOCs.

Despite this yet scarce implementation of gamification in MOOCs, there are several limitations in the current MOOC design for which gamification could be beneficial. High dropout and low completion rates, and a lack of learner engagement are widely presented in the literature as MOOC drawbacks [5]. Gamification could be adopted to increase the level of MOOCs users' engagement, in addition to enabling users to achieve their own goals within a MOOC scenario. Hence, recent studies [6,7] suggest that completion rates alone are not a useful indication of MOOC success, but that participants' goals and intentions also need to be considered [8]. By applying gamification to MOOCs, users could build their own plan and fulfil their personal goals within the course, and the deployment of game elements within MOOC environments could additionally increase the engagement of their users. Games, in particular Massive Multiplayer Online Role Playing Games (MMORPGs), are an example of highly engaging, scalable, though still highly personalised, user experiences [9]. Looking at games can be an option to find those elements that can be transferred to online learning, and more specifically to MOOC environments, with the aim of overcoming the drawbacks highlighted above. However, as mentioned above, there are still only a few examples of empirical studies of gamified MOOCs, thus, in this study, we expanded the investigation by looking also at online learning, which faces similar challenges, i.e., students' dropout [10] and lack of engagement due to the development of feeling of isolation [11]. In this perspective, after having detailed the state-of-the-art of the gamification of MOOCs, the main research questions (Q) of this systematic literature review will be addressed:

1. What are the gamification elements mostly used in online learning environments?
2. What are the effects of these game elements on learners' behaviour in online learning environments?
3. What factors need to be considered for designing effective gamification of online learning environments, specifically MOOCs?

In an attempt to answer these questions, the paper is structured as follows: first, related works are presented, with a focus on the application of gamification of MOOCs as an emerging application field for gamification but still too young to be investigated in this work for its lack of empirical focus, then the method of data collection and analysis of this systematic literature review is detailed, together with the results consisting of the answers to the above reported research questions (Q1 and Q2). The study is rounded off by replying to Q3 and summarising our conclusions, declaring the limitations of this study and introducing our planned future work.

2. Related Works

Gamification effects have been investigated in several domains [12], such as banking [13], trading [14] the medical field [15–17], marketing [18,19] or with the purpose of facilitating annoying tasks [20,21]. According to Dicheva et al. [22], the sector in which gamification has been applied the most is education. In education, several literature reviews have been conducted to describe its state-of-the-art [22–24]; to report on the game elements most used in education [25]; to give an overview on the available gamification design frameworks [26]; to investigate the effects of gamification on students in face-to-face and blended situations [27–30].

Several systematic literature reviews, similar to this study, have been conducted to report on the effects of gamification in online educational settings [31–33]. From these, it is possible to draw the following conclusions: even though the number of publications on the use of gamification in education has increased, the field suffers from (1) a lack of "true empirical research on the effectiveness of incorporating game elements in learning environments" (p. 83), (2) inadequate "methodology used in most of the empirical settings to test the effects of game elements" ([22], p. 83) and (3) "heterogeneous study designs and typically small sample sizes, which highlight the need for further research" ([34], p. 1). However, despite all these *lacunae*, gamification has been judged to be a field with potential, a didactic strategy that if well designed and implemented can really give a boost to users' engagement and motivation.

As far as the investigation of gamification in MOOCs is concerned, several (recent) literature reviews [35–37] report that the field is still developing and lacks empirical studies. Thus, the majority of the papers available are conceptual [6,38–45]. However, by digging into the literature, we found a few studies, and four of these corresponded to our selection criteria (see Table 1). The first is Chang and Wei [46], which represents a first attempt towards the empirical investigation of the gamification of MOOCs. This study focuses on identifying the most suitable game elements to engage students in MOOCs in two steps. Firstly, by conducting a focus group with 25 frequent MOOC users, aimed at collecting the game elements that could trigger the engagement of users in an online setting like an MOOC. The results of this first step was a list of 40 game elements. Secondly, the authors ran a survey aimed at assessing the game elements (40) for their level of engagement (the survey involved 5020 MOOC learners). As a result of the survey data analysis, 10 game elements (virtual goods, three different types of points, leaderboards, trophies and badges, peer grading, emoticon feedback and two types of games) were identified with the highest level of engagement, as chosen by the majority of the survey participants [46]. Unfortunately, we could not find a follow-up study more directed to experimentally validating the hypothesis that these 10 game elements impact MOOC learners' level of engagement.

The second study from the same year is Borras-Gene et al. [47]; they used the platform MiriadaX, for the implementation of their gamified cooperative MOOC (gcMOOC). Several external software programs were used to manage the reinforcement of social aspects of the course, such as Instagram, Google Hangouts and Google+; the game elements (badges) were also not integrated into the platform but issued externally using Mozilla Open Badges. The data collected were both qualitative and quantitative and, in the study, there are not explicit references to A/B testing or the division of participants in gamified cooperative (gc) and plain conditions in the description of the method applied. The conclusions presented are, however, promising for the application of gamification in an environment such as MOOCs.

The third is Binti Mohd Nor Hisham and Sulaiman [48], who conducted an experimental study. The authors [48] deployed the game elements scores, likes (a form of social scoring, similar to the one used in Facebook), progress bars, group activities and rewards, and badges with the purpose of increasing MOOC users' engagement and completion rates. This study involved a total of fifty-three participants, divided in a control group (non-gamified, thirty-one users) and experimental group (gamified, twenty-two users). By comparing the two groups, the results show that there was no significant difference for engagement, but more participants in the experimental group completed the course compared to participants in the control group, therefore gamification impacted positively on completion rates.

The fourth study is from Navío-Marco and Solórzano-García [49], which is a data-driven exploratory study, that involved 3250 participants, investigating how gamification impacted course completion rates. The MOOC was gamified by using badges and 'karma points', a kind of social oriented scoring that works similarly to the 'likes' described above. MOOC participants, instead of assigning 'thumbs up', used these points. In order to determine the effects of their gamified intervention, Navío-Marco and Solórzano-García [49] did not perform an experimental study but via the log data they described and categorised all the participants actions. Their analysis shows that there was a relation between 'karma points' and completion rates: the participants with higher completion rates were the ones with higher 'karma points'.

Based on what has been presented in this paragraph, it is then possible to sum up the following: gamification of MOOCs is a novel field and, from the few empirical studies available, it can be observed that gamification is mainly implemented as a strategy to tackle the MOOCs' drawbacks previously described (completion rate and engagement). The game elements implemented are mainly external rewards, i.e., points, badges and scores. Therefore, a preliminary conclusion could be that the field 'gamification of MOOCs' needs the support of all its researchers and developers to grow and establish empirical evidence. As a consequence, the field needs studies designed, conducted and analysed with rigorous experimental methodologies. Aware of this gap, knowing that gamification is not extensively applied to MOOCs yet, we aim to investigate the effects of gamification in online learning environments and explore if there are some lessons learnt that can be transferred to the MOOC scenario.

3. Methods and Data Collection

This systematic literature review follows the Preferred Reporting Items for Systematic Reviews and Meta-Analyses (PRISMA) [50], represented and described below (Figure 1):

Step 1—Find the right keywords combination

By analysing several literature reviews in the field of Gamification of MOOCs, it became clear to us that there is a lack of empirical works. The few empirical studies available are dated back to very recent years. Therefore, we chose to perform a search aimed at identifying empirical studies on gamified MOOCs and narrow down the time span to the last year (from January 2018 to April 2019). The database used for the search was the university's digital library, which enabled us to simultaneously access several databases: ACM Digital Library, EBSCO, SAGE journals, ScienceDirect (Elsevier), Taylor & Francis Online, Wiley Online Library, Web of Science, Google Scholar, IEEEXplore, Springer Link (and a lot more). The keywords combination applied for this query was *"(gamification) AND ((MOOC*) AND ((experim*) OR (evaluati*))"*, where the first two terms refer to the field of interest, and the other two to the study types we aimed to analyse.

As explained earlier, although our focus is on MOOCs, due to the limited number of empirical studies, this systematic literature review also investigates the effects of gamification applied in online learning environments. Therefore, we have included the terms 'online', 'e-learning' and 'distance', resulting in the following keyword combinations *"(gamification) AND ((MOOC*) OR (online) OR (online) OR (distance) OR (e-learning) OR (education*)) AND ((experim*) OR (evaluati*))"*. In addition, this time we used the university database, filtering the search for the period (January) 2014 to (December) 2018.

Finally, to be sure to gain a high coverage of all available articles in the literature, we have decided to perform a query on those databases that, according to our university library, contained the most records, which were: Google Scholar, IEEE and Web of Science. In this case, we opted for a broader keywords combination: *"Gamification AND Empirical"*, removing any reference in the query to the field of application (i.e., 'MOOC', 'online', 'e-learning', etc.) (the keyword combinations: 'gamification AND effect' and 'gamification AND experiment' were also used, but the results were not satisfying). The period considered was, as above, (January) 2014 to (December) 2018.

Steps 2 to 5

Figure 1 represents and summarises step 2 identification, step 3 screening, step 4 eligibility, and the final step 5, in which the number of papers included in this systematic literature review is identified. To assure the quality of the data analysis and avoid biases due to personal interpretation, the process described in Figure 1 was double-checked by the other two authors of the manuscript.

Inclusion and exclusion criteria for this systematic literature review are detailed in the following Table 1.

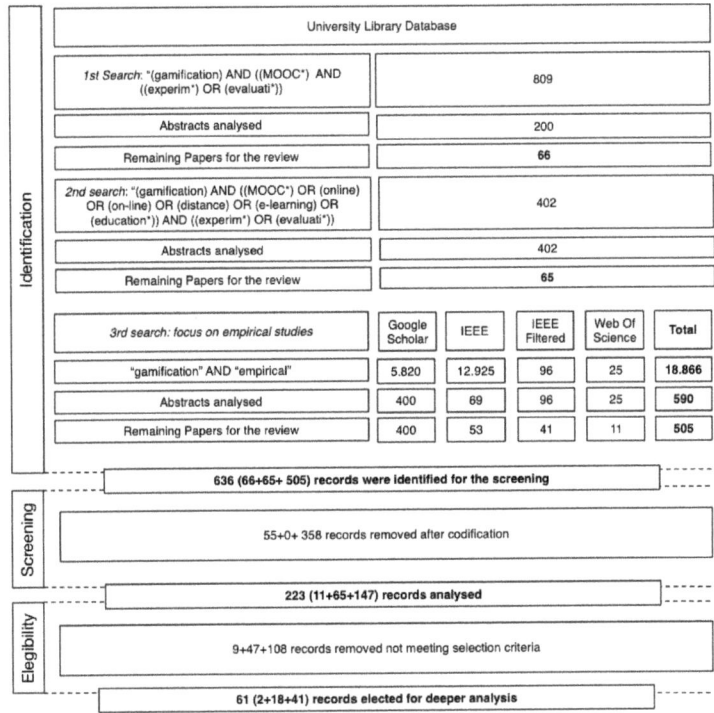

Figure 1. Flow chart of the systematic literature review process according to PRISMA.

Table 1. Criteria for paper selection (steps two to five of the five steps of the systematic literature review performed, omitting the keyword selection).

Criteria	Inclusion Criteria	Exclusion Criteria
Step 2: Identification	• The article refers to empirical studies • The article is related to empirical evidence in the field of gamification and aims to map and/or present the state of the art	• Long abstract only • Thesis • Book • Workshop description • Work-in-progress-paper
Step 3: Screening	• The article belongs to one of the groups: Group 1—Literature Review Group 2—Focus on Online/e-learning/MOOC Group 3—Empirical study but not related to group 2	• Conceptual paper • Not relevant literature review • Paper already found in the other libraries consulted (duplicate) • Not related to gamification but games (education, serious)
Step 4: Eligibility	• The article focuses on game elements and presents data from experimental studies, observational studies, field experiments, quasi experiments • The article combines Online/e-learning/MOOC and empirical studies	• It presents gamification in general • Experiment setting and purpose of the study not in line with the current literature review • It has no focus on gamification but on games • It has no focus on Online/e-learning/MOOC neither present empirical results
Step 5: Included	• After in-depth reading, it really addresses the criteria	• After in-depth reading, it does not really embrace the criteria of selection

4. Data Analysis and Results

The data analysis and results section is structured in accordance with the research questions. Q1 and Q2 will be addressed in the remainder of this section, while Q3 is addressed in the discussion section. A precise overview of the studies, methods used, samples involved and environments, is presented in the Appendix A, Table A1.

4.1. Q1: What Are the Game Elements Most Used in Online Learning Environments?

The lack of empirical studies within MOOCs, and the need to contribute to the field with more awareness of what a single game element or a combination of those can produce as an effect on learners' behaviour, resulted for us in also including in our review studies implemented in e-learning and online learning scenarios (see Appendix A, Table A1). By analysing these studies, we tracked the game elements used: there are 24 (see Appendix A, Table A2), some of them have been studied in isolation, others in combination. More specifically, they are:

1. *Badges*, which are external rewards delivered to users once a goal is accomplished. They have been analysed in 22 studies [47,51–71].
2. *Leaderboards*, a game element based on social comparison that enables a user to understand how s/he is performing (showing the score and/or the position) in relation to others and the leader; this has been adopted and investigated in 15 studies [51–55,58–60,62,63,70,72–75].
3. *Points* that are "a numerical representation of player success" [76]; in the literature, authors also refer to these using the terms 'score' or 'ranking'. They have been examined in 12 studies [52,53,59,61–63,66–69,71,72].
4. *Feedback* can be defined as information delivered to users related to their progress, achievements, issues, or other aspects of their activities. Feedback can take several forms and can be delivered as *direct* or *indirect information*. Sometimes, a clue (information on how to solve a quest) can also be considered feedback. This game element can be used in combination with other game elements, for instance with *leaderboards* and *badges*. We tracked six studies related to feedback [61,66,67,70,71,77].
5. *Challenges* can appear in the form of quizzes or problems to be solved, in *solo* or *team* mode. They can be related to *levels* or/and to *missions*. We recorded the use of this game element in five studies [53–55,66,72].
6. *Likes, social features* are widely spread on social networks, such as Facebook, which implies supporting what another user communicates via a thumbs up (or smileys). Four of the papers found refer to them [47,53,58,61].
7. *Communication Channels* "are the medium and the methods players can use to send messages to other players" [76], an example in this sense are chats. These have been investigated in four studies [51,53,54,61].
8. *Narratives* consist of using stories to pass information and intrigue users, often confused with storytelling, which implies using a character to tell a story. We found that it has been used in four of the studies [67,68,72,78].
9. *Levels* are strictly related to goals and have different degrees of difficulty. To move up a level, it is generally necessary to reach (complete) all the goals of the current level. This game element has been empirically investigated by four studies analysed [54,63,67].
10. *Progress bars* give the user information about his/her own improvement. This game element has been examined by two studies [47,71].
11. *Teams*, referring to working in a team, collaborating with the members within; this was used by two studies [47,72].
12. *Agent*, not to be confused with avatar or profile, is a virtual character by the system (not by the user). In the literature, we have found two studies that implemented agent [71].
13. *Medals*, a form of external reward that have been investigated in two studies [55,69].

14. *Avatar* is the virtual version of the player and has been introduced and investigated in two studies [62,79].
15. *Trophies* are another form of reward for something users have achieved, like solving a level or overcoming a challenge. They are considered in two studies [54,55].
16. *Time limit* considered by [62,70].
17. *Task*, which is generally connected with the previous one, but it has been considered by different authors [79].
18. *Virtual Currency* is a type of reward in the form of virtual money; it has been investigated by [79].
19. *Personalising features*; refers more to features typical of games in which the player can personalise the look and the outfit of the avatar/character; these were discussed in one study [69].
20. *Mission*, which is a type of *challenge*, generally connected with task and time limit, and it has been investigated by [69].
21. *Replayability* is the possibility given to users to re-do an action if, at the first attempt, he/she did not succeed. It was investigated by one study [65].
22. *Goal Indicators* can be combined with several game elements such as levels and missions, and it has been examined by [67].
23. *Competition*, usually within teams and or player vs. player; this has been presented in one study [69].
24. *Win State* very typical of the game world, but it has been considered only by one recent study [72].

4.2. Q2: What Are the Effects of These Game Elements on Learners' Behaviour in Online Learning Environments?

A query of the literature has been done to close the identified gap related to the investigation and definition of what the effects of gamification are on users' behaviour within online environments. From our analysis, we identified mainly six categories of effects of gamification on learners' behaviour and characteristics: (1) performance; (2) motivation; (3) engagement; (4) attitude towards gamification; (5) collaboration; and (6) social awareness. In the following section, the effects per category are described and then summarised in the Appendix A, Table A3 and matched with the game elements in Table A4.

1. Performance

    Grant and Betts [57] examined the influence of badges on (1,295,620) users' performance (activity in web forums); the positive impact of badges on users' performance were investigated by Hakulinen et al. [58]'s study, who investigated badges, within an online learning environment (involving 281 students) and their impact on the carefulness of the performance (looking at significant differences in terms of time spent per submission and number of attempts per exercise), highlighting the fact that badges can affect students' behaviour in terms of time needed to perform and precision in the performance increasing students' carefulness. "Students in the treatment group spent more time per submission on average" (p. 23), showing a higher level of carefulness compared to their colleagues in the control group [58]. In a similar scenario, an online learning course (involving 150 students), Kyewski and Krämer [64] investigated the effects of badges on learners' performance, and, in contrast to the studies previously detailed, recorded a negative effect of badges on learners' performance.

    Leaderboards were the focus of studies by Landers and Landers [75] and Bernik et al. [51], both implemented in an online learning scenario. The first (involving a total of 86 participants) aimed to study the effect of the game element leaderboard on learners' academic performance, looking at the amount of time spent on a task (consisting of a wiki project). As a result, users in the treatment group (with leaderboard) increased their time-on-task performance [75]. The second, Bernik et al. [51] (involving a total of 55 students), looked at performance comparing the achievements of the control and treatment groups. Students in the treatment group had much better achievements, when comparing the average results in the post-tests performed by the two groups [51].

The single game element narrative was studied by Armstrong and Landers [78], who tested how users would perform and react to gamified training, finding that participants in the gamified condition reacted better to the training than their colleagues in the control (non-gamified) condition [78].

Several studies have investigated the effects of gamification on users' performance by combining different game elements. Four game elements: challenges, leaderboards, trophies and medals were implemented by Domínguez et al. [55] within an e-learning platform. To study the effects of their plugin on students' learning performance, they compared the scores achieved in several activities by participants in both control (non-gamified) and treatment (gamified) groups. The data analysis showed that students in the treatment group performed better than those in the control group in the initial activity, and in the majority of the assignments. Negative effects of the gamification approach were instead reported for the final examination, where students in the treatment group registered a significantly lower score [55]. Tsay et al. [70] also opted for four game elements: immediate feedback; badges; leaderboard and time limits. By comparing the scores of the two groups (control-non-gamified and treatment-gamified), the authors highlighted that students in the gamified condition performed better (higher score) than their colleagues in the control group. Finding that their performance was mediated by their level of engagement, the students in the gamified condition were more engaged and therefore had better performance compared to their colleagues in the control condition [70].

The study of De-Marcos et al. [79] aimed to determine which of these five different conditions: plain, educational game, gamification, social networking and social, was the most effective on learning performance (75 undergraduate students were involved). Participants, distributed into the different conditions, were asked to perform five different tests (word processing, spreadsheets, presentation, databases and final examination). In the comparative analysis, it was demonstrated that all the experimental conditions outperformed the control group in three of these tests (word processing, spreadsheets and presentation). The authors concluded that gamification impacts on learning performance when it is combined with social approaches [79]. Similar conclusions were drawn by Krause et al. [62], who showed that social gamification increases learning success (see details of this study in the engagement section).

Long and Aleven [65] studied the effect of a gamified Intelligent Tutoring System (ITS) on students' learning and enjoyment (involving 267 students). The game elements selected were: rewards (Rwd) for good performance; and replay (RePl), the chance to complete a task again (in this case a problem already solved). The four different conditions were analysed combining the two game elements (RePl + Rwd; No-RePl + Rwd; RePl + No-Rwd; No-RePl + No-Rwd) and the most effective combination for students' learning was replay without rewards (RePl + No-Rwd) [65]. Students under the condition with both replay and rewards activated, performed significantly worse than their colleagues exposed to the other three conditions of the study. No significant effects were found when comparing the two groups in terms of enjoyment [65].

Two studies recorded the effects of gamification on performance in terms of goal commitments. The first by Landers et al. [74] reported on a quasi-experiment (involving 339 university students), to better understand how leaderboards can enhance user performance and influence goal commitment, or vice versa [74]. Participants were randomly assigned to one of five conditions (do-your-best, easy goal, difficult goal, impossible goal and leaderboard). With their study, Landers et al. [74] demonstrated that the leaderboard condition was the most influential and that the goal-setting theory is suitable to understand and explain leaderboard effects on users' performance: "commitment moderates the success of leaderboards as goal-setting theory would predict" ([74], p. 5). In accordance with the goal-setting theory [80], goal commitment and performance are directly related to each other, "goals are only effective if people are committed to them, and performance is maximised when individuals are committed to difficult, specific goals" ([74], p. 5). The results showed that users in the leaderboard condition performed better than those in the other conditions. Participants in the leaderboard condition were "likely to target the top or near-top goals presented on that leaderboard, even without specific instructions to target those goals" ([74], p. 6). The second study, conducted by Hakulinen and Auvinen [81], focused more on goal orientation. The authors studied the relationship between the achievement of badges and goal orientations, they hypothesised "that students with different goal orientation profiles respond differently to badges" ([81], p. 9). Their study was conducted in an online environment (involving 278 students) divided into a control (non-gamified) and a treatment (gamified) group. They "found no statistically significant differences in the behaviour of the different goal orientation groups regarding badges. However, their attitudes towards the badges varied" ([81], p. 16).

2. Motivation

The effect of rewards in the form of badges on learner motivation was studied by Kyewski and Krämer [64]. They first measured the participants' level of the intrinsic motivation before the intervention. Subsequently, they checked: (1) if the intrinsic motivation level changed in relation to participants earning the badge; (2) if the level of intrinsic motivation changed compared to the beginning (higher intrinsic motivation for those who had a low level at the beginning and vice-versa, lower for those who declared a high level of intrinsic motivation). No effects of badges were found on students' intrinsic motivation. These results contradicted a study conducted in similar conditions (participants from the university level, working in an online learning environment) by Hakulinen et al. [58], reporting that the majority of participants declared being motivated by badges and concluding that "achievement badges seem to be a promising method to motivate students" ([58], p. 18). The positive effect of badges on learners' motivation was confirmed by Gooch et al. [56], who conducted their study in Moodle but with a small group of children with a learning disability (dyslexia). Based on their findings, badges significantly improved students' motivation [56].

Dyslexic students (40 were involved) were also the target audience of Saputra [67]'s study, who adopted the following game elements: narrative, goal indicators, levels, scores, feedback and badges, highlighting that these game elements produce positive effects on students' engagement, enjoyment and motivation in the short term. A longitudinal study instead was conducted by Hanus and Fox [60] to assess the effects of badges and leaderboards on students' motivation, as well as on social comparison, effort, satisfaction, learner empowerment, and academic performance. Two classes (gamified and not-gamified) were set up and the results showed that, over time, students in the gamified course presented less motivation, satisfaction and empowerment compared to those involved in the non-gamified class [60].

The study of Utomo and Santoso [71] was implemented in an online learning environment to test the effects of a 'pedagogical agent' on students' motivation and behaviour towards learning. The agent activity was enhanced with the adaptation of some game elements, such as progress bar, badges, scores and feedback. Based on students' evaluation, the authors concluded that personalised feedback in real time boosted learners' motivation "toward active learning behaviour" ([71], p. 7). Badges, as well as points and leaderboards, were used also by Huang and Hew [59], who gamified an SPSS (Statistical Package for Social Sciences) course in 'Moodle', showing that badges and leaderboards motivated most of the learners. Points engaged students and stimulated them to undertake challenging tasks and extracurricular learning, while "learners in the control group did not attempt any challenge" [59]. The same game elements were implemented in LOPUPA (Learning On Projects of United Promotion for Academia), a gamified platform proposed by Kuo and Chuang [63] implemented with the purpose of engaging and motivating online academic dissemination. Based on the data collected via the platform (as well as google analytics), and the feedback given by participants by filling in a questionnaire, the authors showed increased engagement and motivation of the participants that used the platform [63]. Similar conclusions were reached by Domínguez et al. [55] who, in their study, also investigated the effect of their gamification design, described above, on students' motivation, stating that "gamification in e-learning platforms seems to have the potential to increase student motivation, but it is not trivial to achieve that effect, and a big effort is required in the design and implementation of the experience for it to be fully motivating for participants" (p. 391).

3. *Engagement*

Kyewski and Krämer [64] found that earning badges in an online learning environment did not increase students' level of activity, which was used as a measure of engagement. Furthermore, data "revealed that students in the no-badge condition who did not earn badges were even more active than students in the gamification conditions" ([64], p. 32). This contradicts previous research findings, such as those reported by Sitra et al. [68], suggesting positive effects of badges on students' engagement, even though they tested it within a different population (small number of students with special needs), using an LMS system-Moodle similar to Kyewski and Krämer [64]. Badges as well as narrative, score/ranking, levels, and quests are covered in the study of [69]. They aimed at investigating whether these game elements affect learners' motivation and engagement on a peer assessment platform. By comparing the results of control and treatment groups, they found that the number of elaborated essays as well as the number of corrected essays were higher for participants in the treatment group. Therefore, the data indicated that gamification stimulated students to use the platform more, thus enhancing engagement [69].

The following three studies underlined how the 'social factor' is important for engaging students. Firstly, Krause et al. [62] set up three conditions in 'Moodle' (plain–no game elements, game–with game elements, social-with social game elements). In particular, the game condition consisted of implementing, in an online environment (Moodle), game elements such as: avatars, badges, points, leaderboards and time limits. The social-game condition included all the game elements listed above and in addition integrated social game elements encouraging competition among students, also in remote mode, via pre-recorded actions, mainly by using the leaderboard and mechanism of social comparison. The purpose was to understand the differences (if any) between gamification and social gamification on students' retention and learning success. The data showed that learners in the social game condition were more engaged compared with their colleagues in the plain condition, and the social game elements enhanced gamification effects on retention and success [62]. Secondly, De-Marcos et al. [54] set up a quasi-experimental design with five conditions (control group, non-gamified, gamified, social non-use, social use) with the purpose of understanding the effects of 'social networking' and 'gamification' on students' academic achievement, engagement and attitude in an undergraduate course [54]. The game elements implemented were: badges, challenges, leaderboards, levels, trophies and forum. The data revealed that both gamification and social networking were perceived positively by students, therefore an improvement of participants' attitude towards these approaches was recorded, while no significant statistical differences were found in students' academic achievements or engagement [54]. Thirdly, Mazarakis [77] conducted a study on using feedback mechanisms, testing four types of feedback to increase user engagement, i.e., participation in the course wiki: (1) gratitude feedback: this expressed thankfulness without giving any further information to the users, (2) historical: this gave information about users' activities and contributions, (3) 'relative ranking': showing relative user rank, and (4) 'social ranking feedback': this aimed at providing information about the points and activated competition via ranking comparison. The results showed that providing feedback mechanisms can enhance participation. Among the feedback mechanisms tested, the one enabling social comparison (called 'social ranking feedback') was the most effective feedback [77]. More focused on the attitude towards gamification is the study of Aldemir et al. [72], included in this dimension because it deals also with the aim of increasing students' perception of others within an online educational scenario. The study investigated in particular the participants' attitude towards leaderboards, challenges, narratives, teams, badges (and rewards in general) win-states, points, and 'constraints', providing generally positive results and "insights about game elements integrated into a gamified course in both online and face-to-face sessions" ([72], p. 251).

4. *Attitude towards gamification*

The following authors: Bernik et al. [51], Hakulinen et al. [58], Aldemir et al. [72], whose work is already presented above, have also all studied the effects of their gamification designs on the attitude towards gamification, reporting all positive responses of the participants in this matter. In the same line is the study of [54], who in an online learning scenario set up a quasi-experiment with the purpose of studying the effect of social networking and gamification on student academic achievement, participation and attitude towards gamification. The participants were split into three conditions: gamification (114); social network (185) and control (75). In the gamified condition, the following game elements were displayed: badges, challenges, leaderboards, forum, levels and trophies. A survey was completed by both the experimental groups to assess learners' attitudes towards gamification and social networks. The results showed a positive effect on users' attitude towards gamification; no significant effects were found on performance improvement and neither on written examinations. Regarding academic achievement, it was better for the social network participants [54].

5. *Collaboration*

Knutas et al. [61] studied the effects of a gamified online discussion system on users' collaborative behaviour and communication (involving 249 students). The system was used during an existing university course 'introduction to programming' for fourteen weeks. The major aim of the gamified online discussion system was to facilitate contributions for effective discussions. Therefore, to enhance peers' contribution and communication, users were enabled to 'like' or 'unlike' the comments of their peers. In addition, a reward feedback system was implemented for the users who were contributing. To determine the effects of the described gamified online discussion system on users' collaborative behaviour and communication, several tools were used, such as surveys, interviews and Social Network Analysis (SNA). From the data analysis, it was possible to conclude that the gamified online discussion system increased student collaboration and course communication efficiency by reducing response time to students' questions. In particular, the survey showed that skilled students liked the gamification features (consisting of the discussion system and rewards) and Knutas et al. [61] concluded that the game elements stimulated users to contribute more by giving more answers and proposing more questions. Part of this dimension is also in the Aldemir et al. [72] study, described above, who point out that, in using the team game element, the balance of "the team skills" is important, enabling the right level of competition and collaboration within the team.

6. *Social awareness*

Two studies were dedicated to the "social awareness" that gamification can generate: Aldemir et al. [72], Christy and Fox [73]. In particular, [73] set up a study on leaderboards, aimed at investigating whether interaction with a leaderboard (with male vs. female leaders) produced effects in the social comparison condition or led to stereotype threats and effects on academic performance. Their data showed that "leaderboards appear to have inspired social comparison processes" (p. 74) more than 'stereotype threats' [73]. The second study by [72] adopted "teams", another game element that can enable participants, in an online learning course, to be aware of the others and thus facilitate community building. Using the authors' words: "Community-building process is affected by the interaction and relationship between teammates, implying that good communication facilitates community building. Another implication is that, the fewer people in a team, the easier it is to communicate; therefore, the teams should be small community building" ([72], p. 250). Based on the qualitative data collected by Aldemir et al. [72], participants reacted positively to this game element, with positive effects on learning achievement.

Several studies have analysed moderating factors of gamification effects, specifically considering gender, personality and lifestyle (work) of the participants. Pedro et al. [66]; Tsay et al. [70]; and Codish and Ravid [52] studied respectively whether gamification had different effects on female and male students; on participants with different lifestyles (part-time or full-time job); and on extroverts vs. introverts. It turned out that, according to Pedro et al. [66], males are more influenced by gamification (in particular by eternal rewards, such as: points, badges, feedback, scores and challenges) compared to females. However, the situation changes if the population works part-time, indeed according to the results of Tsay et al. [70]: "The gamified course [...] was particularly beneficial to learners who were working part-time and female students" ([70], p. 32), while, according to Codish and Ravid [52], extroverts connect playfulness with badges, rewards and points while introverts tend to prefer an offline leaderboard.

5. Discussion

In this systematic literature review, we have investigated and reported on the effects that gamification can generate on human behaviour, within MOOCs and online learning environments. Three research questions have been addressed.

Q1 relates to game elements used most in online learning environments, which are, in order of frequency of usage: badges/rewards (most used); leaderboards and points/score/ranking. Addressing Q2, we clustered the effects of these game elements in six dimensions: performance, motivation, engagement, attitude towards gamification, collaboration and communication, and social awareness. We highlighted the application scenarios and characteristics of each study by underlining what type of effects a specific game element or combination could generate on users.

Both the game elements examined in the studies included in this systematic literature review and their effects have been matched and are presented in the Appendix A, Table A4. By correlating the game elements most used in online learning and the effects they generate on learners, it is possible to notice the following:

- Effects of *badges/rewards* are observed on motivation, attitude toward gamification use, and performance in terms of time management, engagement, emotional states, and enjoyment. Effects of badges may vary according to gender and personality, and, if perceived as controlling and restrictive, they may negatively affect motivation and engagement. They can be used to set clear goals or to stimulate social comparison, both variables that have a positive effect on performance (in the training field, specifically).
- Effects of *leaderboards* have been found on attitude toward gamification use, learning performance, performance in general, engagement, enjoyment, and goal commitment, by engaging students in difficult tasks. Effects generated by leaderboards vary according to personality. Just like badges, leaderboards can enable social comparison that can positively influence performance. More specifically, the game element leaderboards provides information about points, scored by users, which activates social comparison and competition among them and achieve, as effects, higher participation and engagement, in particular at the cognitive level [77].

- Effects of *point/score/ranking* have been reported related to motivation, attitude toward gamification use, learning performance, performance in general, engagement, enjoyment and emotional states. Effects of point/score/ranking may vary according to personality and gender. They can foster social comparison and thus also influence performance. Furthermore, the game element points/scores/ranking enhanced the level of users' engagement so much that they were stimulated to undertake challenging tasks and target "top or near-top goals" in terms of difficulties [74].

In relation to the lessons learnt from the application of gamification to online leaning environments, transferrable to MOOCs and, therefore, Q3, the following considerations are made. First, the gamification of online learning environment design addresses similar problems that MOOCs have: motivation, performance, and engagement. Second, from the studies found, it is possible to highlight the multidisciplinary character of gamification and its design. The third and most important lesson learnt is that each game element "should be carefully chosen [...] to avoid gamification from being just unnecessary eye candy" ([81], p. 28). As in the MOOCs, field external rewards are largely used in online learning. While, on one side, they are easy to be implemented, it is important to keep in mind that their meaning and purpose can be easily misunderstood by the receivers Kyewski and Krämer [64]. Hanus and Fox [60] explained that, when an audience is already motivated, the reward strategy might be perceived negatively, "interpreted as controlling, causing students to feel less confident, to be less satisfied with the course, and to have less motivation to engage with the material" ([60], p. 159). This result is in line with the Cognitive Evaluation Theory (CET) [82], which predicts that external events can harm intrinsic motivation "based on whether individuals process those events as informational or controlling" ([60], p. 153). Therefore, if a reward, such as a badge, is used as mere 'candy for the eyes', then gamification could be far from beneficial.

Summing up, the following factors need to be considered in designing gamification for MOOCs: A first "vital aspect of gamification design is the context of application" ([83], p. 3). In the framework of this study, the application scenario is MOOCs. MOOCs can appear in several formats, but they all have a common denominator: a massive audience, which implies an enormous differentiation in users' culture, age, prior knowledge, background, languages, intentions, etc., all factors that need to be taken into consideration in designing an intervention within such an environment. The second factor is the problem, which determines the aim of the intervention: it can differ from case to case (and can be, for instance, related to lack of personalisation, of engagement, high user dropout rate, etc.). The third factor is the selection and the design of the most appropriate game elements in accordance with the problem to be addressed and the desired effect/s. In this case, the definition of a theoretical framework that supports the design of your game elements, as well as your approach will be of added value.

MOOCs are online learning scenarios, with their own peculiarities. These are due essentially to the massive audience and the technology used (platform) and its *openness*, implying with openness the degree of manoeuvre that designers and developers have in directly intervening on the platform, which is a factor that can drastically impact on the gamification design choices, the effectiveness of the implementation and the intervention in an MOOC scenario.

Last but not least, it is important to also consider the evaluation phase when designing the gamification of MOOCs (or other application scenarios). By conducting rigorous research in the field of gamification applied to online learning environments, the awareness about its potential and limits can be increased and the field can grow further.

6. Future Work

This systematic literature review had the general purpose of contributing to the field by increasing awareness about the empirically tested effects of gamification on human behaviour in online learning environments. To raise this awareness, the state-of-the-art in the 'Gamification of MOOCs' has been presented. Subsequently, the game elements most used in online learning scenarios (Q1) have been identified, and their effects on users' behaviours (Q2). Then, the lesson learnt from online learning has been transferred to MOOCs and the steps needed to design gamification of MOOCs have been drawn up (Q3).

Our conclusion and analysis is in line with Nacke and Deterding [83], who underline that "gamification research is maturing ([83], p. 4), and there have been steps forward aiming at understanding the effects of gamification; however, "many studies are still to some extent comparing apples with oranges, testing different implementations of design elements with different effect measures" (p. 3). In relation to the design of gamification, other game design elements are introduced but still PBL (points badges and leaderboards) remain the favourite.

In our future work, we aim to conduct rigorous research with the aim of, from one side, contributing to the field with a framework on how to design proper gamification Antonaci et al. [42] and, from the other, on how to study the effects of our design on human behaviour, contributing to the growth of the field with empirical data and studies. Furthermore, we aim in future work to demonstrate that gamification can be applied in a more sophisticated way than just PBL, a gamified way that can be easily confused with a simple 'candy for the eyes', and an external reward system that does not consider the variegated characteristics that an MOOC audience can have. More specifically, in our future studies, we aim to design an intervention, using gamification to tackle the problems we have identified in MOOCs (lack of engagement and goal achievement of MOOC users). From the literature, we could deduce that, to enhance engagement within an online learning scenario, the 'social aspect' plays a strategic role. In a recent study, Tseng et al. [84] analysed the *social presence* in games and the factors that make game communities powerful. In addition, they also underlined the factors that retain players in a specific group or 'guilds', which are: firstly, *sense of community* (membership, sense of belonging) that positively impacts retention and 'relation switching cost' [84]; secondly, *interdependence* that can be described as the "degree to which members in a community rely on each other to make decisions and take actions" ([84], p. 603). Hence, as future work, to address the lack of engagement of MOOC users, we aim to identify game elements that can generate a sense of community and interdependence. Our hypothesis is: *if MOOC users generate a sense of community, then their level of engagement and retention will increase*.

As far as the enhancement of MOOC users' goal achievement via gamification is concerned, in previous studies, we have identified the Implementation Intention theory by Gollwitzer [85] as promising for designing our game element/s. This theory shows that, if a person makes a precise plan (detailing the *when*, *what* and *how* of reaching her/his own goal), taking also into account what to do if an inconvenience occurs, the chance of reaching her/his goal/s will be higher compared with people that only have an intention [86]. Our hypothesis is *if MOOC users have the chance to make a plan, according to the implementation intention theory, then their goal achievement will be higher compared with users that do not make a plan*. Considering that, in this study, none of the studied game elements enable MOOC users to develop a sense of community or implement an intention, we will expand our analysis looking at game elements collections such as Björk and Holopainen [76] and type of games such as Massive Multiplayer Online Role Playing Games (MMORPGs) that might match our needs. Once identified, we will validate the game element selection involving experts from the fields of game design, pedagogy and instructional design. In the end, an experimental study will be set up based on our hypothesis to test, analyse and validate the empirical effects of all our selected game elements on engagement (sense of community) and goal achievement of our MOOC learners.

7. Limitations of This Study

Several shortcomings can be identified with two major limitations: firstly, the fact that a very limited number of studies were available in the literature with a specific focus on gamification of MOOCs. Secondly, to map the effect of a specific game element in a certain scenario, a meta-analytic approach would have been more suitable. However, due to the diversity in the studies, the sample and the methodologies applied, it was not possible to conduct a meta-analysis that could determine more precisely the effects of each game element in an online scenario.

Author Contributions: Conceptualization, A.A., R.K. and M.S.; the methodology used was agreed upon by the three authors; formal analysis, resources, investigation, data curation and writing—original draft preparation, A.A.; writing—review and editing, R.K. and M.S.; visualization and validation, A.A.; supervision, R.K. and M.S. For term explanation, check http://img.mdpi.org/data/contributor-role-instruction.pdf.

Funding: This research received no external funding.

Acknowledgments: This research has been conducted in the framework of a PhD study, partially founded by Erasmus+ projects.

Conflicts of Interest: The authors declare no conflict of interest. The funders had no role in the design of the study; in the collection, analyses, or interpretation of data; in the writing of the manuscript, or in the decision to publish the results.

Abbreviations

The following abbreviations are used in this manuscript:

MOOCs	Massive Open Online Courses
MMORPGs	Massive Multiplayer Online Role Playing Games
Q	Research Question
MMOGs	Massive Multiplayer Online Games

Appendix A

Table A1. Methods used in the studies included in this systematic literature review, with details about sample and environment gamified.

Study	Method	Sample	Environment
Bernik et al. [51]	Pre-post survey	55 (28 + 27)	e-learning
De-Marcos et al. [53]	Experimental	75	e-learning
De-Marcos et al. [54]	Quasi-exp	318 (62 + 111 + 146)	e-learning
Domínguez et al. [55]	Quasi-exp	173 (62 + 111)	e-learning
Borras-Gene et al. [47]	Mixed Methods	3866	MiriadaX
Gooch et al. [56]	Pre-post tests	22 (10 + 10 + 2)	ClassDojo
Armstrong and Landers [78]	Experimental (two courses compared)	273	e-learning
Hakulinen et al. [58]	Experimental	281	online learning
Christy and Fox [73]	Experimental	76	online
Aldemir et al. [72]	Empirical (observation, interviews, documents)	118	e-learning
Huang and Hew [59]	Quasi-exp	40 (21 + 19)	online learning
Knutas et al. [61]	Empirical (SNA, survey, interviews)	249	online learning
Krause et al. [62]	Experimental	206 (71 + 67 + 68)	Moodle
Kuo and Chuang [63]	Empirical (surveys, web analytics)	73	LOPUPA
Kyewski and Krämer [64]	Experimental	126	Moodle
Long and Aleven [65]	Experimental	190	online learning
Utomo and Santoso [71]	Empirical (questionnaire and focus groups)	31	Moodle
Tsay et al. [70]	Empirical (two versions of the course compared)	136	Moodle
Pedro et al. [66]	Empirical (questionnaire)	16	E-Game
Saputra [67]	Empirical (observation and questionnaire)	40	e-learning
Sitra et al. [68]	Case Study	5	Moodle
Tenorio et al. [69]	Experimental	32	MeuTutor
Mazarakis [77]	Experimental	436	online learning
Landers et al. [74]	Experimental	240	online learning
Landers and Landers [75]	Experimental	109	online learning
Codish and Ravid [52]	Empirical (questionnaire)	102 (58 + 44)	online learning
Grant and Betts [57]	Experimental	1,295,620	Stack Overflow

Table A2. Map of the game elements most used in online learning environments.

Game Elements	No. of Studies	References
Badges/Rewards	22	[47,51–71]
Leaderboard	15	[51–55,58–60,62,63,70,72–75]
Points/Score/Ranking	12	[52,53,59,61–63,66–69,71,72]
Feedback	6	[61,66,67,70,71,77]
Challenge	5	[53–55,66,72]
Likes (social features)	4	[47,53,58,61]
Communication Channels	4	[51,53,54,61]
Narrative	4	[67,68,72,78]
Levels	3	[54,63,67]
Progress Bar	2	[47,71]
Teams	2	[47,72]
Agent	2	[71]
Medals	2	[55,69]
Avatar	2	[62,79]
Trophies	2	[54,55]
Time Limit	2	[62,70]
Task	1	[79]
Virtual Currency	1	[53]
Personalising Features	1	[79]
Missions	1	[69]
Replayability	1	[65]
Goal indicators	1	[67]
Competition	1	[69]
Win State	1	[72]

Table A3. The users' behaviour and characteristics affected by gamification.

Cluster	Effects	References
Performance	Performance and time management	[51,53,55,57,58,62,64,65,70,75,78,81]
	Performance and goal commitment	[74,75,81]
Motivation		[47,55,56,58–60,63,64,66,67,69,71]
Engagement	Engagement	[51,58,59,61,63,64,68–70,77]
	Retention	[62]
	Enjoyment	[52,65,67,68,78]
Attitude towards gamification		[47,51,54,58,72]
Collaboration (and communication)		[61,72]
Social awareness	Community building	[72]
	Social comparison	[73]

Table A4. The game elements applied in online learning and their effects matched.

Areas Impacted—Game Elements	Engagement	Engagement (Retention)	Enjoyment	Performance Commitment (Goal Commitment)	Attitude towards Gamification	Motivation	Collaboration (Communication)	Community Building	Social Comparison
Badges	[51,58,59,61,63,64,68–71]	[62]	[65,67,68]	[51,53,57,58,62,70]	[47,51,54,58]	[47,55,56,58–60,64,66,67,69]	[61]		
Leaderboard	[59,63,70,72]	[62]		[52,53,62,70,72,74,75,81]	[51,54]	[55,59,60]			[73]
Points/scores	[59,61,63,69,72]	[62]	[68]	[51,53,62,71,72]	[51]	[52,59,66,67,69]	[61]		
Feedback	[61,71,77]		[67]			[66,67,70]	[61]		
Challenges				[53,55,66]	[54,72]	[55]	[72]		
Likes (social features)	[53,61]				[47,58]				
Communication Channels	[63]			[51,53]	[51,54]	[63]	[62]		
Narrative			[67,68,78]		[72]	[67]			
Levels	[63]		[67]		[54]	[67]			
Progress bar	[71]								
Profile Pages	[51]			[51,53]			[47]		
Teams					[72]		[47]	[72]	
Agent						[71]			
Medals	[69]					[69]			
Avatar		[62]		[62]					
Trophies			[65]	[53]	[54]				
Time Limits		[62]		[62]					
Task			[67]	[53]	[54]				
Virtual currency				[53]					
Personalising Features	[53]	[54]							
Mission	[69]					[69]			
Replayability			[65]	[65]					
Goal Indicators						[67]			
Competition	[69]					[69]			
Win State							[72]		

References

1. Kalz, M.; Specht, M. *If MOOCS Are the Answer, Did We Ask the Right Questions? Implications for the Design of Large-Scale Online-Courses*; Maastricht School of Management: Maastricht, The Netherlands, 2013.
2. Kizilcec, R.F.; Piech, C.; Schneider, E. Deconstructing disengagement: analyzing learner subpopulations in massive open online courses. In Proceedings of the Third International Conference on Learning Analytics and Knowledge, Leuven, Belgium, 8–13 April 2013.
3. Pappano, L. The Year of the MOOC. *The New York Times*, 14 November 2012.
4. Deterding, S.; Dixon, D.; Khaled, R.; Nacke, L. From game design elements to gamefulness: Defining "gamification". In Proceedings of the 15th International Academic MindTrek Conference on Envisioning Future Media Environments (MindTrek '11), Tampere, Finland, 28–30 September 2011; ACM: New York, NY, USA, 2011; pp. 9–15. [CrossRef]
5. Reich, J.; Ruipérez-Valiente, J.A. The MOOC pivot. *Science* **2019**, *363*, 130–131. [CrossRef] [PubMed]
6. Antonaci, A.; Klemke, R.; Stracke, C.M.; Specht, M. Gamification in MOOCs to enhance users' goal achievement. In Proceedings of the 2017 IEEE Global Engineering Education Conference (EDUCON), Athens, Greece, 25–28 April 2017; pp. 1654–1662. [CrossRef]
7. Henderikx, M.A.; Kreijns, K.; Kalz, M. Refining success and dropout in massive open online courses based on the intention–behavior gap. *Distance Educ.* **2017**, *38*, 353–368. [CrossRef]
8. Reich, J. MOOC completion and retention in the context of student intent. *EDUCAUSE Review Online*, 8 December 2014.
9. Kihl, M.; Aurelius, A.; Lagerstedt, C. Analysis of World of Warcraft traffic patterns and user behavior. In Proceedings of the International Congress on Ultra Modern Telecommunications and Control Systems, Moscow, Russia, 18–20 October 2010; pp. 218–223. [CrossRef]
10. Drachsler, H.; Kalz, M. The MOOC and learning analytics innovation cycle (MOLAC): A reflective summary of ongoing research and its challenges. *J. Comput. Assist. Learn.* **2016**, *32*, 281–290. [CrossRef]
11. Croft, N.; Dalton, A.; Grant, M. Overcoming isolation in distance learning: Building a learning community through time and space. *J. Educ. Built Environ.* **2010**, *5*, 27–64. [CrossRef]
12. Kasurinen, J.; Knutas, A. Publication trends in gamification: A systematic mapping study. *Comput. Sci. Rev.* **2018**, *27*, 33–44. [CrossRef]
13. Rodrigues, L.F.; Oliveira, A.; Costa, C.J. Playing seriously—How gamification and social cues influence bank customers to use gamified e-business applications. *Comput. Hum. Behav.* **2016**, *63*, 392–407. [CrossRef]
14. Hamari, J. Do badges increase user activity? A field experiment on the effects of gamification. *Comput. Hum. Behav.* **2015**, 1–10. [CrossRef]
15. Marques, R.; Gregório, J.; Pinheiro, F.; Póvoa, P.; Da Silva, M.M.; Lapão, L.V. How can information systems provide support to nurses' hand hygiene performance? Using gamification and indoor location to improve hand hygiene awareness and reduce hospital infections. *BMC Med. Inform. Decis. Mak.* **2017**, *17*, 1–16. [CrossRef]
16. Ryan, J.; Edney, S.; Maher, C. Engagement, compliance and retention with a gamified online social networking physical activity intervention. *Transl. Behav. Med.* **2017**, *7*, 702–708. [CrossRef]
17. Sardi, L.; Idri, A.; Fernández-Alemán, J.L. A systematic review of gamification in e-Health. *J. Biomed. Inform.* **2017**, *71*, 31–48. [CrossRef]
18. Kim, K.; Ahn, S.J.G. The Role of Gamification in Enhancing Intrinsic Motivation to Use a Loyalty Program. *J. Interact. Mark.* **2017**, *40*, 41–51. [CrossRef]
19. Yang, Y.; Asaad, Y.; Dwivedi, Y. Examining the impact of gamification on intention of engagement and brand attitude in the marketing context. *Comput. Hum. Behav.* **2017**, *73*, 459–469. [CrossRef]
20. Mekler, E.D.; Brühlmann, F.; Opwis, K.; Tuch, A.N. Do points, levels and leaderboards harm intrinsic motivation? In Proceedings of the First, International Conference on Gameful Design, Research, and Applications (Gamification '13), Toronto, ON, Canada, 2–4 October 2013; pp. 66–73.
21. Mekler, E.D.; Brühlmann, F.; Tuch, A.N.; Opwis, K. Towards understanding the effects of individual gamification elements on intrinsic motivation and performance. *Comput. Hum. Behav.* **2015**, 1–10. [CrossRef]
22. Dicheva, D.; Dichev, C.; Agre, G.; Angelova, G. Gamification in Education: A Systematic Mapping Study. *Educ. Technol. Soc.* **2015**, *18*, 75–88. [CrossRef]

23. De Sousa Borges, S.; Durelli, V.H.S.; Reis, H.M.; Isotani, S. A Systematic Mapping on Gamification Applied to Education. In Proceedings of the 29th Annual ACM Symposium on Applied Computing (SAC '14), Gyeongju, Korea, 24–28 March 2014; ACM: New York, NY, USA, 2014; pp. 216–222. [CrossRef]
24. Caponetto, I.; Earp, J.; Ott, M. Gamification and Education: A Literature Review. In Proceedings of the European Conference on Games Based Learning, Berlin, Germany, 9–10 October 2014; Volume 1, pp. 50–57.
25. Nah, F.F.H.; Zeng, Q.; Telaprolu, V.R.; Ayyappa, A.P.; Eschenbrenner, B. Gamification of Education: A Review of Literature. In *International Conference on HCI in Business*; Nah, F.F., Ed.; Springer: Cham, Switzerland, 2014; pp. 401–409.
26. Mora, A.; Riera, D.; Gonzalez, C.; Arnedo-Moreno, J. A Literature Review of Gamification Design Frameworks. In Proceedings of the 7th International Conference on Games and Virtual Worlds for Serious Applications (VS-Games), Skövde, Sweden, 16–18 September 2015; pp. 1–8. [CrossRef]
27. Çakıroğlu, Ü.; Başıbüyük, B.; Güler, M.; Atabay, M.; Yılmaz Memiş, B. Gamifying an ICT course: Influences on engagement and academic performance. *Comput. Hum. Behav.* **2017**, *69*, 98–107. [CrossRef]
28. De Almeida Souza, M.R.; Constantino, K.F.; Veado, L.F.; Figueiredo, E.M.L. Gamification in Software Engineering Education: An Empirical Study. In Proceedings of the IEEE 30th Conference on Software Engineering Education and Training (CSEE&T), Savannah, Georgia, 7–9 November 2017; pp. 276–284. [CrossRef]
29. Monterrat, B.; Lavoué, É.; George, S. Adaptation of Gaming Features for Motivating Learners. *Simul. Gaming* **2017**, *48*, 625–656. [CrossRef]
30. Smith, T. Gamified Modules for an Introductory Statistics Course and Their Impact on Attitudes and Learning. *Simul. Gaming* **2017**, *48*, 832–854. [CrossRef]
31. Hamari, J.; Koivisto, J.; Sarsa, H. Does gamification work?—A literature review of empirical studies on gamification. In Proceedings of the Annual Hawaii International Conference on System Sciences, Waikoloa, HI, USA, 6–9 January 2014; pp. 3025–3034. [CrossRef]
32. Lister, M.C. Gamification: The effect on student motivation and performance at the post-secondary level. *Issues Trends Educ. Technol.* **2015**, *3*, 1–22. [CrossRef]
33. Looyestyn, J.; Kernot, J.; Boshoff, K.; Ryan, J.; Edney, S.; Maher, C. Does gamification increase engagement with online programs? A systematic review *PLoS ONE* **2017**, *12*, e0173403. [CrossRef]
34. Lumsden, J.; Edwards, E.A.; Lawrence, N.S.; Coyle, D.; Munafò, M.R. Gamification of Cognitive Assessment and Cognitive Training: A Systematic Review of Applications and Efficacy. *JMIR Serious Games* **2016**, *4*, 1–14. [CrossRef]
35. Khalil, M.; Wong, J.; Ebner, M.; de Koning, B.; Ebner, M.; Paas, F. Gamification in MOOCs: A Review of the State of the Art. In Proceedings of the IEEE Global Engineering Education Conference, Santa Cruz de Tenerife, Spain, 18–20 April 2018; pp. 1635–1644. [CrossRef]
36. Ortega-Arranz, A.; Muñoz-Cristóbal, J.A.; Martínez-Monés, A.; Bote-Lorenzo, M.L.; Asensio-Pérez, J.I. How gamification is being implemented in MOOCs? A systematic literature review. In *European Conference on Technology Enhanced Learning*; Lavoué, É., Drachsler, H., Verbert, K., Broisin, J., Pérez-Sanagustín, M., Eds.; Springer: Tallinn, Estonia, 2017; Volume 10474, pp. 441–447.
37. Bakar, N.F.A.; Yusof, A.F.; Iahad, N.A.; Ahmad, N. The Implementation of Gamification in Massive Open Online Courses (MOOC) Platform. In *International Conference on User Science and Engineering*; Springer: Singapore, 2018; pp. 183–193.
38. Gené, O.B.; Mart, M.; Blanco, Á.F. Gamification in MOOC: Challenges, Opportunities and Proposals for Advancing MOOC Model. In Proceedings of the 2nd International Conference on Technological Ecosystems for Enhancing Multiculturality, Salamanca, Spain, 1–3 October 2014; ACM: Salamanca, Spain, 2014; pp. 215–220. [CrossRef]
39. Saraguro-Bravo, R.A.; Jara-Roa, D.I.; Agila-Palacios, M. Techno-Instructional Application in an MOOC Designed with Gamification Techniques. In Proceedings of the Third International Conference on eDemocracy & eGovernment (ICEDEG), Quito, Ecuador, 30 March–1 April 2016; pp. 176–179. [CrossRef]
40. Mesquita, M.A.A.; Toda, A.M.; Brancher, J.D. BrasilEduca—An Open-Source MOOC platform for Portuguese speakers with gamification concepts. *IEEE Front. Educ. Conf.* **2014**, 446–449. [CrossRef]
41. Vaibhav, A.; Gupta, P. Gamification of MOOCs for increasing user engagement. In Proceedings of the IEEE International Conference on MOOCs, Innovation and Technology in Education (IEEE MITE 2014), Patiala, India, 19–20 December 2014; pp. 290–295. [CrossRef]

42. Antonaci, A.; Klemke, R.; Kreijns, K.; Specht, M. Get Gamification of MOOC right! How to Embed the Individual and Social Aspects of MOOCs in Gamification Design. *Int. J. Serious Games* **2018**, *5*. [CrossRef]
43. Antonaci, A.; Peter, D.; Klemke, R.; Bruysten, T.; Christian, M.; Specht, M. gMOOCs—Flow and Persuasion to Gamify MOOCs. In *International Conference on Games and Learning Alliance*; Lecture Notes in Computer Science (LNCS) Series; Springer: Cham, Switzerland, 2017; Volume 10653, pp. 126–136.
44. Antonaci, A.; Klemke, R.; Stracke, C.M.; Specht, M. Identifying game elements suitable for MOOCs. In *European Conference on Technology Enhanced Learning*; Springer: Cham, Switzerland, 2017; pp. 355–360.
45. Antonaci, A.; Klemke, R.; Stracke, C.M.; Specht, M. Towards Implementing Gamification in MOOCs. In *International Conference on Games and Learning Alliance*; Lecture Notes in Computer Science (LNCS) Series; Springer: Cham, Switzerland 2017; Volume 10653, pp. 115–125.
46. Chang, J.W.; Wei, H.Y. Exploring engaging gamification mechanics in massive online open courses. *Educ. Technol. Soc.* **2016**, *19*, 177–203.
47. Borras-Gene, O.; Martinez-Nuñez, M.; Fidalgo-Blanco, Á. New Challenges for the Motivation and Learning in Engineering Education Using Gamification in MOOC. *Int. J. Eng. Educ.* **2016**, *32*, 501–512.
48. Binti Mohd Nor Hisham, F.; Sulaiman, S. Adapting Gamification Approach in Massive Open Online Courses to Improve User Engagement. In *UTM Computing Proceedings Innovation in Computing Technology and Applications*; UTM: Skudai, Malaysia, 2017; Volume 2, pp. 1–6.
49. Navío-Marco, J.; Solórzano-García, M. Student's social e-reputation ("karma") as motivational factor in MOOC learning. *Interact. Learn. Environ.* **2019**, 1–15. [CrossRef]
50. Moher, D.; Liberati, A.; Tetzlaff, J.; Altman, D.G.; PRISMA Group. Preferred Reporting Items for Systematic Reviews and Meta-Analyses: The PRISMA Statement. *PLoS Med.* **2009**, *6*, e1000097. [CrossRef]
51. Bernik, A.; Bubaš, G.; Radoševi, D. A Pilot Study of the Influence of Gamification on the Effectiveness of an e-Learning Course. In *Central European Conference on Information and Intelligent Systems*; Faculty of Organization and Informatics Varazdin: Varazdin, Croatia, 2015; pp. 73–79.
52. Codish, D.; Ravid, G. Personality Based Gamification—Educational Gamification for Extroverts and Introverts. In *CHAIS '14—Conference for the Study of Innovation and Learning Technologies: Learning in the Technological Era*; Eshet-Alkalai, Y., Caspi, A., Geri, N., Kalman, Y., Silber-Varod, V., Yair, Y., Eds.; The Open University of Israel: Raanana, Israel, 2014; pp. 36–44.
53. De-Marcos, L.; García-López, E.; García-Cabot, A.; Medina-Merodio, J.A.; Domínguez, A.; Martínez-Herraíz, J.J.; Diez-Folledo, T. Social network analysis of a gamified e-learning course: Small-world phenomenon and network metrics as predictors of academic performance. *Comput. Hum. Behav.* **2016**, *60*, 312–321. [CrossRef]
54. De-Marcos, L.; Domínguez, A.; Saenz-De-Navarrete, J.; Pagés, C. An empirical study comparing gamification and social networking on e-learning. *Comput. Educ.* **2014**, *75*, 82–91. [CrossRef]
55. Domínguez, A.; Saenz-De-Navarrete, J.; De-Marcos, L.; Fernández-Sanz, L.; Pagés, C.; Martínez-Herráiz, J.J. Gamifying learning experiences: Practical implications and outcomes. *Comput. Educ.* **2013**, *63*, 380–392. [CrossRef]
56. Gooch, D.; Vasalou, A.; Benton, L.; Khaled, R. Using Gamification to Motivate Students with Dyslexia. In *Proceedings of the CHI Conference on Human Factors in Computing Systems (CHI '16)*, San Jose, CA, USA, 7–12 May 2016; ACM: New York, NY, USA, 2016; pp. 969–980. [CrossRef]
57. Grant, S.; Betts, B. Encouraging user behaviour with achievements: An empirical study. In Proceedings of the IEEE International Working Conference on Mining Software Repositories, San Francisco, CA, USA, 18–19 May 2013; pp. 65–68. [CrossRef]
58. Hakulinen, L.; Auvinen, T.; Korhonen, A. The effect of achievement badges on students' behavior: An empirical study in a university-level computer science course. *Int. J. Emerg. Technol. Learn.* **2015**, *10*, 18–30. [CrossRef]
59. Huang, B.; Hew, K.F. Do points, badges and leaderboard increase learning and activity: A quasi-experiment on the effects of gamification. In Proceedings of the 23rd International Conference on Computers in Education, Hangzhou, China, 30 November–4 December 2015; Society for Computer in Education: Hangzhou, China, 2015; pp. 275–280.
60. Hanus, M.D.; Fox, J. Assessing the effects of gamification in the classroom: A longitudinal study on intrinsic motivation, social comparison, satisfaction, effort, and academic performance. *Comput. Educ.* **2015**, *80*, 152–161. [CrossRef]

61. Knutas, A.; Ikonen, J.; Nikula, U.; Porras, J. Increasing collaborative communications in a programming course with gamification. In Proceedings of the 15th International Conference on Computer Systems and Technologies (CompSysTech '14), Ruse, Bulgaria, 27–28 June 2014; Volume 883, pp. 370–377. [CrossRef]
62. Krause, M.; Mogalle, M.; Pohl, H.; Williams, J.J. A Playful Game Changer: Fostering Student Retention in Online Education with Social Gamification. In Proceedings of the 2nd ACM Conference on Learning@Scale (L@S'15), Vancouver, BC, Canada, 14–18 March 2015; pp. 95–102. [CrossRef]
63. Kuo, M.S.; Chuang, T.Y. How gamification motivates visits and engagement for online academic dissemination—An empirical study. *Comput. Hum. Behav.* **2016**, *55*, 16–27. [CrossRef]
64. Kyewski, E.; Krämer, N.C. To gamify or not to gamify? An experimental field study of the influence of badges on motivation, activity, and performance in an online learning course. *Comput. Educ.* **2018**, *118*, 25–37. [CrossRef]
65. Long, Y.; Aleven, V. Gamification of Joint Student / System Control over Problem Selection in a Linear Equation Tutor. In *International Conference on Intelligent Tutoring Systems*; Trausan-Matu, S., Boyer, K.E., Crosby, M., Panourgia, K., Eds.; Lecture Notes in Computer Science (LNCS) Series; Springer: New York, NY, USA, 2014; pp. 378–387.
66. Pedro, L.Z.; Lopes, A.M.Z.; Prates, B.G.; Vassileva, J.; Isotani, S. Does gamification work for boys and girls? In Proceedings of the 30th Annual ACM Symposium on Applied Computing—SAC '15, Salamanca, Spain, 1–17 April 2015; ACM: Salamanca, Spain, 2015; pp. 214–219.
67. Saputra, M.R.U. LexiPal: Design, Implementation and Evaluation of Gamification on Learning Application for Dyslexia. *Int. J. Comput. Appl.* **2015**, *131*, 37–43. [CrossRef]
68. Sitra, O.; Katsigiannakis, V.; Karagiannidis, C.; Mavropoulou, S. The effect of badges on the engagement of students with special educational needs: A case study. *Educ. Inf. Technol.* **2017**, *22*, 3037–3046. [CrossRef]
69. Tenorio, T.; Bittencourt, I.I.; Isotani, S.; Pedro, A.; Ospina, P. A gamified peer assessment model for online learning environments in a competitive context. *Comput. Hum. Behav.* **2016**, *64*, 247–263. [CrossRef]
70. Tsay, C.H.H.; Kofinas, A.; Luo, J. Enhancing student learning experience with technology-mediated gamification: An empirical study. *Comput. Educ.* **2018**, *121*, 1–17. [CrossRef]
71. Utomo, A.Y.; Santoso, H.B. Development of gamification-enriched pedagogical agent for e-Learning system based on community of inquiry. In Proceedings of the International HCI and UX Conference in Indonesia on (CHIuXiD '15), Bandung, Indonesia, 8–10 April 2015; pp. 1–9. [CrossRef]
72. Aldemir, T.; Celik, B.; Kaplan, G. A qualitative investigation of student perceptions of game elements in a gamified course. *Comput. Hum. Behav.* **2018**, *78*, 235–254. [CrossRef]
73. Christy, K.R.; Fox, J. Leaderboards in a virtual classroom: A test of stereotype threat and social comparison explanations for women's math performance. *Comput. Educ.* **2014**, *78*, 66–77. [CrossRef]
74. Landers, R.N.; Bauer, K.N.; Callan, R.C. Gamification of task performance with leaderboards: A goal setting experiment. *Comput. Hum. Behav.* **2015**, 1–8. [CrossRef]
75. Landers, R.N.; Landers, A.K. An Empirical Test of the Theory of Gamified Learning: The Effect of Leaderboards on Time-on-Task and Academic Performance. *Simul. Gaming* **2014**, *45*, 769–785. [CrossRef]
76. Björk, S.; Holopainen, J. *Patterns in Game Design*; Charles River Media, Inc.: Needham, MA, USA, 2005.
77. Mazarakis, A. Using Gamification for Technology Enhanced Learning: The Case of Feedback Mechanisms. *Bull. IEEE Tech. Comm. Learn. Technol.* **2015**, *17*, 6–9.
78. Armstrong, M.B.; Landers, R.N. An Evaluation of Gamified Training: Using Narrative to Improve Reactions and Learning. *Simul. Gaming* **2017**, *48*, 513–538. [CrossRef]
79. De-Marcos, L.; Garcia-Lopez, E.; Garcia-Cabot, A. On the effectiveness of game-like and social approaches in learning: Comparing educational gaming, gamification & social networking. *Comput. Educ.* **2016**, *95*, 99–113. [CrossRef]
80. Locke, E.A.; Latham, G.P. New Directions in Goal-Setting Theory. *Curr. Dir. Psychol. Sci.* **2006**, *15*, 265–268. [CrossRef]
81. Hakulinen, L.; Auvinen, T. The effect of gamification on students with different achievement goal orientations. In Proceedings of the International Conference on Teaching and Learning in Computing and Engineering (LATICE 2014), Kuching, Malaysia, 11–13 April 2014; pp. 9–16. [CrossRef]
82. Ryan, R.M.; Deci, E.L. Self-determination theory and the facilitation of intrinsic motivation, social development, and well-being. *Am. Psychol.* **2000**, *55*, 68–78. [CrossRef]

83. Nacke, L.E.; Deterding, S. The maturing of gamification research. *Comput. Hum. Behav.* **2017**, *71*, 450–454. [CrossRef]
84. Tseng, F.C.; Huang, H.C.; Teng, C.I. How Do Online Game Communities Retain Gamers? Social Presence and Social Capital Perspectives. *J. Comput.-Mediat. Commun.* **2015**, *20*, 601–614. [CrossRef]
85. Gollwitzer, P.M. Implementation intentions. *Am. Psychol.* **1999**, *54*, 493–503. [CrossRef]
86. Gollwitzer, P.M.; Sheeran, P. Implementation intentions and goal achievement: A meta-analysis of effects and processes. *Adv. Exp. Soc. Psychol.* **2006**, *38*, 69–119. [CrossRef]

© 2019 by the authors. Licensee MDPI, Basel, Switzerland. This article is an open access article distributed under the terms and conditions of the Creative Commons Attribution (CC BY) license (http://creativecommons.org/licenses/by/4.0/).

Article

Company–University Collaboration in Applying Gamification to Learning about Insurance

Teresa Rojo [1,*], Myriam González-Limón [2] and Asunción Rodríguez-Ramos [3]

1   Departamento de Sociología, Universidad de Sevilla, 41001 Sevilla, Spain
2   Departamento de Análisis Económico y Economía Política, Universidad de Sevilla, 41001 Sevilla, Spain; miryam@us.es
3   Departamento de Economía e Historia Económica, Universidad de Sevilla, 41001 Sevilla, Spain; asunrod@us.es
*   Correspondence: trojo@us.es; Tel.: +34-6-5042-2068

Received: 16 July 2019; Accepted: 5 September 2019; Published: 19 September 2019

Abstract: Incorporating gamification into training–learning at universities is hampered by a shortage of quality, adapted educational video games. Large companies are leading in the creation of educational video games for their internal training or to enhance their public image and universities can benefit from collaborating. The aim of this research is to evaluate, both objectively and subjectively, the potential of the simulation game BugaMAP (developed by the MAPFRE Foundation) for university teaching about insurance. To this end, we have assessed both the game itself and the experience of using the game as perceived by 142 economics students from various degree plans and courses at the University of Seville during the 2017–2018 academic year. As a methodology, a checklist of gamification components is used for the objective evaluation, and an opinion questionnaire on the game experience is used for the subjective evaluation. Among the results several findings stand out. One is the high satisfaction of the students with the knowledge acquired using fun and social interaction. Another is that the role of the university professors and the company monitors turns out to be very active and necessary during the game-learning sessions. Finally, in addition to the benefits to the university of occasionally available quality games to accelerate student skills training, the company–university collaboration serves as a trial and refinement of innovative tools for game-based learning.

Keywords: gamification; university; education; serious video games; game-based learning; professors; video games design; knowledge; skills training; digital technologies; automated learning

---

1. Introduction

Gamification is one of the great technological advances of the last decades. It has been developed mostly inside the video game industry in an effort to improve the attractiveness of its entertainment products. Recently, such know how is being transferred to a wide variety of sectors such as education and training, health and fitness, transport, service-marketing, environment, corporate management, welfare initiatives, politics, etc. [1,2], with the purpose of designing gamified experiences, simulation games, or serious video games that will increase their activity performance. In 2018, the video game industry turnover amounted to $137.9 billion worldwide, which is higher than the volume of the music business. Of this amount, 87% is derived from digital businesses via the internet [3]. The number of players in the world is already 2.3 billion, 47% of whom are women. The growth rate of the sector is 8% per year with serious video games having an increasing weight in that growth.

Gamification applied to university teaching is the focus of our study. Gamification can be defined, specifically when applied to the education sector, as a system of procedures for the flexible application of game playing elements to skills development and/or knowledge learning. It follows the line of the generally accepted definition proposed by Djeerling, Dixon, Khaleed, and Nacke: "Gamification, as the use of game design elements in non-game contexts" ([4], p. 9). The term gamification is also used when just one or very few game design elements are applied without producing a game as such. That was the focus of some research on gamification in Hamari's work on the impact of "badges" in customer engagement [5] or the study on gamification in education by Dicheva, Dichev, Agre, and Angelova who excluded games as such [6]. Besides, the debate on the categories for classifying, either game design elements or implementation impacts, remains quite open because gamification brings together multidisciplinary aspects where game design technology and human–computer interaction (HCI) increasingly share this space with social sciences like psychology, pedagogy, marketing, business, or sociology [4]. This multiplies the number of approaches and scientific production. The data offered by Hamari, Kolvisto, and Sarsa on the number of empirical studies on gamification published in SCOPUS journals or registered in Google Scholar show an increase from an average of 50 in 2011 to 160 in 2012 and 200 in 2013 [1].

Regarding the education sector, a revolutionary technological innovation in the field of learning such as gamification, is very attractive to teachers dealing with students who are products of a rapidly developed digital culture. Faculty members, particularly at the secondary and university levels, face new generations of students who are part of the digital culture, and the video game genre opens new perspectives for accelerating experiential learning. Serious simulation video games, such as those for learning how to fly planes or perform surgery, have already had great commercial success in the education sector, due to the savings represented by the virtual formula of acquiring practical skills while avoiding the real risks of possible destruction or injury, resulting from wrong decisions.

But those teachers who have an interest in transferring gamification to their classes encounter various types of difficulties. First and foremost are the financial difficulties due to the high cost of designing a video game of sufficient quality to attract the interest of students. In addition, teachers lack training in the areas of systems analysis and design of educational video games and, therefore, are not able to collaborate effectively with video game manufacturing companies. A study on gamification in education concludes that early adopters of gamification in universities tend to be the computer science/information technology educators; limited by the lack of appropriate supportive technology [6]. Another obstacle that stands out is the attitude of the parents of the students and those responsible for the educational policy who tend to express an opinion of the video game as a distraction instead of an instrument for accelerated learning.

This introduction presents a selection of contributions of state-of-the-art and controversial interpretations concerning: The trends of the video game industry towards the production of serious video games for education; the gamification elements that make a product more attractive to users or consumers; the attitudes shown by the public sector; and university teaching staff towards gamification as an instrument or set of techniques for learning. The introduction ends with an explanation of the aim of this study: Evaluating company–university collaboration in applying gamification to university education. To this end, we use a case study involving the participation of groups of economics students from the University of Seville in the 2017–2018 university championship of the BugaMAP game (MAPFRE Foundation), a simulation game used for successfully managing insurance companies.

1.1. The Production of Serious Video Games in the Video Game Industry

In the video game industry there is a moderate production of serious video games. In Spain, for example, 14% of all video games produced annually are serious video games [7]. Among serious games there are those for the education sector, health, culture, public administration, marketing, or others.

Furthermore, it tends to be the big companies, large Non-Governmental Organizations (NGOs) and international organizations, and not so much the universities, that are leading the production of serious video games and likewise assuming the costs of making gamified learning available to the citizens and the educational sector ([8], p. 665; [9], p. 1053).

The video game industry is the leader in gamification. It dominates a global market industry, with multi-billion-dollar business figures ($137 billion in 2017) and it reaches an audience of some 2.3 billion players with products in several languages. Of every 10 dollars of the video game market, 5 are spent in the Asia Pacific region, 3 in the American continent, and 2 in Europe, Middle East, and Africa regions.

By country, the main video game consumption markets are China and the US (with $38 and $30 billion, respectively). They are followed by Japan with about $20 billion. And then come countries such as South Korea, Germany, the United Kingdom, France, Canada, and Spain with markets of $4 to $2 billion. Spain, with 46 million inhabitants and 36 million people connected online, is ranked 9th in the world market of video games by country with a volume of $2.03 billion.

The industry was born in the 1970s and one of its major technological breakdowns occurred in 2004–2005 when free software became available for the design of video games and it also became possible to play online without having to buy games. Thus, companies and products grew exponentially, especially "indie" (independent) companies that would upload their games to platforms to be tested by the users. In the last decade, access to games through mobile devices led to this way of gaming being preferred by 95% of the world's players by 2018 [3].

In 2007, the year of the iPhone, the video game industry was a $35 billion industry, and in just one decade it grew to $137 billion. Nowadays the video game market is classified as those who play (playing), those who watch videos about games (viewing), and those who buy them (owning). According to the CEO of Newzoo Consultancy, a new large expansion of the industry is coming due to the demand for gamification from traditional sectors [3].

> "As we look towards the future, we foresee games playing an increasingly impactful role in disrupting and reshaping traditional industries, from implementing individual game mechanics to an array of mergers and acquisitions crossing the boundaries of traditional industries." ([3], p. 5)

Other authors have offered glimpses into the past regarding the expansion of the serious games market. Michaud et al. ([10], p. 7) estimated in 2010 that global revenue from serious games was $1.5 billion, and that it would have a projected sales growth rate of 47% between 2010 and 2015. Other sources such as Markets and Markets (2015) [11] estimated that the serious games market could reach $5.5 billion in 2020.

To identify the institutions that lead in the production of serious video games, the case of serious environmental video games has been studied because information about them has already been analyzed by Rojo and Dudu (2017) [12]. Their review of previous studies (see below) shows that for the North American market, the prominent producers of serious video games are large companies, universities, and independent developers. In Spain, the leading producers are large companies and NGOs, followed by regional public administrations.

That conclusion for the American market came from a study by Katsaliaki and Musafee (2015) ([8], p. 665) who analyzed 49 sustainable development video games. As for the study on serious environmental video games in Spain, Ouriachi et al. ([9], p.1053) analyzed 24 such games for young people which were accessible online free of charge. In comparing both studies, the conclusion is that universities play an important role in serious games production in the North American market, but they are not currently involved in Spain.

1.2. Gamification as a Unique Way to Engage in Learning

In the presentation of the 2018 Newzoo video game industry report, its CEO and co-creator Peter Warman, acknowledged that its strong growth and success was due to "our industry's ability to offer unique ways to engage" ([3], p. 3). The way to engage is the main achievement of gamification as a technological innovation that has developed within the video game industry from its beginnings in 1970 to the present. And, according to authors such as Tom Chatfield [13], gamification consists of creating a virtual environment in which activities get the participant hooked by using seven different ways of rewarding the human brain (see Table 1).

Table 1. Gamification brain reward techniques for a pleasant learning experience. Source: elaborated from Tom Chatfield 2010.

1. Levels of experience that measure progress	2. Multiple short- and long-term objectives
3. Gifts for efforts	4. Answering in a quick, frequent, and clear way
5. Uncertain final outcome. Increases dopamine	6. Windows with information, both to help decision making and to learn from decisions impact
7. Other persons (collective game or comparing and comments with others)	

Thus, gamification manages to make learning experiences pleasant, or to seduce users towards simulated behaviors that are rewarding to their brains. And among the gamification tools pointed out by Chatfield (2010) [13] in Table 1, several can be highlighted: setting clear objectives, rapid and frequent response, timely contributions of helpful information, the element of uncertainty about the final results, and contact with other players.

French researcher Elisabeth Grimaud (2017) [14] uses the term "dose" to refer to brain neurotransmitters that favor brain synapses. Dopamine, for example, is mobilized with the pursuit of goals and the success of achieving them. Oxytocin is associated with qualities of affection; thus, it is mobilized by the encounter with others. Serotonin is what the brain produces when we are happy, proud, and satisfied with ourselves. Endorphins occur in the management of the effort to continue, to keep going. So, the information that comes to us, or the behaviors that a gamified learning process asks us to perform, can achieve these types of rewarding brain chemical changes.

Lara Boyd (2015) [15] studied the effects of video games on the recovery of patients with cerebral infarction. She found that changes in the brain are both functional and structural when patients practice behaviors to achieve learning skills or abilities. The first good news is that the brain learns throughout life for the neuroplasticity that characterizes it. The second is that the best driver of neuroplastic change in the brain is behavior.

As behaviors are practiced to develop skills or abilities, the brain changes functionally. Whether one is a musician, taxi driver, researcher, or bricklayer, each profession with its own skills and abilities usually activates certain springs of the brain. In that way, when a capacity or ability has developed sufficiently, a part or region of the brain becomes accustomed to being activated until it develops more.

But, mastering an activity can require around 10,000 h of practice. This is where gamification intervenes and accelerates learning by creating practical experience levels of virtual behavior. Faced with the challenges such as curbing climate change and reducing pollution by means of resource-saving behaviors, serious environmental video games contribute to accelerate citizen responses (Rojo and Dudu, 2017 [16]). Also, simulation games involving risk behaviors, such as piloting planes or performing surgery, are accelerating learning and reducing costs in time and equipment for the training of these professionals.

Other investigations such as the framing theory, modernly reformulated by authors such as Lakoff (2010) [17], reinforce the importance in gamification of "setting clear objectives" and "providing information windows with clear, fast and frequent responses" on the part of the game. According to Lakoff (2010) [17], the theory of framing proposes that communicating information or data on an issue is only effective when the recipient has a framework or system of neurological frames in the brain, in which the data or information can fit or articulate with one's accumulated knowledge. Thus, ideas framed in a certain way are settling in the brain.

1.3. Gifts for Effort and Motivating through Gamification

In the gamification of virtual experiences, the first clear idea that designers need to have is the direction in which the person is motivated, what behaviors are desired or most valued in the game or gamified experience. One can differentiate two modes of motivation in the content of a game: intrinsic and extrinsic.

(a) Intrinsic motivation is usually different for each game/person. It is about what attracts the person towards a task or a set of tasks that result in achieving and living certain experiences. And, there will be a range of activities/tasks such that the user likes some more than others. Authors such as Kapp et al. ([18], p. 164) advise designing this intrinsic motivation aspect of the game through its content in narrative, its history, or the mystery or puzzle to be solved such that the action itself is the prize. This means looking for design components that lead the player or user to live the experience. According to experts in school motivation, Ryan and Deci (2000) [19], intrinsic motivation results in high quality learning and creativity; it accrues from engaging in interesting activities and then behaviors become "volitional and accompanied by experience of freedom and autonomy" ([19], p. 65).

Werbach and Hunter (2012) [20] establish three basic rules that must be fulfilled for an activity to intrinsically motivate a person: competence, relation, and autonomy. According to Ryan and Deci (2000) [19], those three social context rules are relevant for intrinsic motivation because they support three innate needs the person faces when he/she is exposed to new ideas and exercises new skills: to feel connected, effective, and agentic ([19], p. 65):

- Competence—that the person can be effective in the tasks performed. According to Frasca (2012) [21], gamification has the effect of activating the desire to improve personal performance in a task. A final "try again" message for example, makes it possible for the user to make mistakes without dramatic consequences. According to Ryan and Deci's research on education, if the student feels efficacious, he/she will more easily adopt a goal as their own, thus in order to facilitate internalization of a goal they recommend offering "optimal challenges and effectance-relevant feedback" ([19], p. 64).
- Relation—that the person can interact and be involved. To describe this aspect, Smethurst (2015) [22] suggests using "interreacting with the game" as the most appropriate term to emphasize that it consists of an innovation that goes beyond simply "interacting" which was already possible with more traditional media such as television or radio. In their self determination theory about motivation in school (SDT), Ryan and Deci name this aspect "sense of relatedness" meaning "sense of belongingness and connectedness to the persons, group, or culture" which is obtained by being valued by significant others to whom one feels connected.
- And, autonomy—that the person can feel in control of his/her own life. That is, the user makes decisions and discovers his/her own way (more on the aspect of decisions below). According to research done by various authors reviewed by Ryan and Deci ([19], p. 63) autonomy and self-regulation contexts facilitate internalization of goal and behavioral regulations at schools. By the contrary, when students were more externally regulated, they showed a tendency to blame others for negative outcomes while more autonomy correlated with interest and enjoyment ([19], p. 63).

(b) Extrinsic motivation in gamification systems consists of dynamizing through scores, prizes, and certificates. Scores or points inform the player of progress, and work as a continuous evaluation that is very stimulating. Prizes or cards (badges) as well as certifications recognize intermediate successes and degrees reached (leaderboards), motivating the player towards enhanced performance. In their review of literature on gamification of education, Nah, Telaprolu, Ayyappa and Eschenbrenner agree to the above mentioned modalities of gifts and add two more: level/stages and progress bar (Nah et al. 2014 [23]).

Kapp et al. ([18], p. 147) recommend a procedure for rewarding the player. First establish the needs and objectives of the game, then determine what should be measured and then give those components a value. Points measure mainly the dedication of the player and progression in the tasks. They provide explicit and frequent feedback. In many video games, points are redeemable for possessions or provide access to resources needed for better performance. It can also be seen in terms of goal orientation and "layers" of goals, where reaching a long-term goal requires having completed medium-term and short-term goals, as well as missions broken into multiple tasks (Nah et al. 2014 [23]).

The prizes (badges) are usually applied when a certain score is reached, often in the form of gradients that are achieved within a ladder. Hamari ([5], p. 476) studied the effect of applying a gamification element such as batch system to the case of Sharetribe, an international peer-to-peer trading service. Its results show that customer engagement efficiency tends to decrease when it ceases to be a novelty and that the consumer's personality could be a significant variable to explain the differences in behavior when faced with the batch system.

Badges give the player a rank in terms of level reached in one or several competitions, and engenders empathy with those who have reached similar ranges. They are reputation markers. "Leaderboards" offer a final qualification in the game by making public the results of all the players and making it possible to compare oneself to others.

Werbach and Hunter ([20], p. 76) point out some aspects that can be demotivating, such as games with "leaderboards" that remain static, games that reward activities that are intrinsically motivating, or games that exhibit conflicts between prizes and achievements.

1.4. Other Aspects in Gamification: Decisions and Narrative

From the analysis of intrinsic motivation, two new fundamental elements stand out for the success of a gamification process: narrative and decision making. So, it seems convenient to add them to Chatfield's list. Narrative refers to the fact that the story, experience, or activity proposed in the game or game experience has interest in itself. Ryan and Deci (2000) [19] refer to intrinsic interest activities as those that "have the appeal of novelty, challenge, or aesthetic value for the individual".

Generally, it is well known that those are aspects that anyone looks for to submerse in a good story with epic moments. The narrative structure refers to the script of the game itself and what happens in the game in terms of the process that is followed to experience the behavior the game is designed to train. As Werbach and Hunter (2012) [20] suggest, the narrative is about defining a magic circle, a little world that is meaningful to the players. The game is what happens inside that circle.

Also, within the gamification there is a prominent role for creating options and decision-making opportunities. The game has to offer elections. The sense of control that players have in games is powerful because of the choices they make that produce an emotion of dominance, of power, over the events and results of the game. Not all games are fun, but they are volunteers. The one who plays makes decisions, chooses, with consequences that affect the gaming experience.

For authors such as Cervera (2012) [24], discussed below, "make own decisions and take on the consequences that they have" as well as "be yourself and live your own experience" are two of the important skills acquired and trained by the video game player. From the perspective of game design or simulation, to train those skills you have to create decision making options and a story, a "narrative".

1.5. Trends towards Gamification in University Education to Train Generic and Transversal Skills

Institutional public discourse to promote gamified learning in public education is rare. Only some international institutions such as United Nations Educational, Scientific, and Cultural Organization (UNESCO) and the United Nations (UN), openly express support for educational video games, promote them for training on issues of peace and environmental protection, and even support expert recommendations on their design [25].

Within the public education sectors, however, the number of professors who favor the incorporation of gamification into the courses they teach is increasing. Contreras and Eguía (2016) [26] argue that traditional education is perceived by many students as boring and that teachers face the challenge of motivating students who are digital natives. There are basically two schools of thought among professors. There are those who believe that video games will replace them as workers, and there are those who believe that video games require supervision by the professor and bring greater productivity to their work.

Players with more than 10 years of experience with video games such as H. Cervera (2012) [24], recognize that they have acquired skills and competences that include: taking risks, having purposes, thinking long term, managing resources, making their own decisions and taking the consequences, and learning to collaborate with others (see Table 2). Many of these skills are specific competences or generic transversal competences that are part of the curricula of university degrees.

Table 2. Gamification and skills acquired by the video game player. Source: Humberto Cervera (2012) [24].

1. Have purposes that get results	2. Think long-term
3. Manage and optimize resources	4. Take risks, dare to risks
5. Compete in face of achievable challenges	6. Ask all the possible questions
7. Be yourself or live your own experiences	8. Make your own decisions and take on the consequences that they have
9. Explore the surroundings	10. Collaborate with others in multiplayer games, learn to play together.

For Contreras and Eguía (2016) [26], the main skills that video games should teach are, in addition to the acquisition of knowledge, the resolution of problems (all skills in Table 2 are involved in problem resolution), and collaboration, or communication (both are included in skill 10 in Table 2). The skills listed in Table 2 are reinforced by the conclusions of the video game designer McGonigal (2015) [27], and creator of the game "superbetter". McGonigal especially points out three groups of psychological skills or strengths that the games help a person to develop:

- The ability to control attention: thoughts and feelings (skills 1 and 2 in Table 2);
- The power to convert others into potential allies and to be able to strengthen the relationships that we already have (skill 10 in Table 2);
- Natural ability to motivate and overload heroic qualities (willpower, compassion, and determination) (skills 4, 5, 7, 8, and 9 in Table 2).

This means that the skills that video games usually train have a high equivalence with those pursued in universities. Therefore, the implementation of gamification in university education, whether through games or gamified experiences, is an unavoidable trend for these avant-garde institutions. But we have also seen that the political environment and cultural attitudes are somehow adverse to expect that in the short term universities achieve financial autonomy to undertake gamification projects or quality video games on their own.

On the contrary, we have seen that private companies in different branches of the economy do show a state of alert and progress in the production of serious video games or in the gamification of their products and services. The video game companies themselves have also visualized the expansion of the gamification market and are becoming active.

In this situation, how can universities advance in the gamification of their teachings? To what extent can universities use collaboration modalities with companies that own quality games to share their products or for joint developments? In order to answer the questions, we have studied the case of the collaboration of the Universidad de Sevilla (Spain) with the Foundation of the MAPFRE Insurance Company, the BugaMAP game owners (see Figure 1), in order to implement the game with economics students of the 2018–2019 academic year. Until then, the game had only been used by the MAPFRE Foundation for internal training of its employees, as well as in universities in Latin America [28].

In BugaMAP simulation, students in each session are distributed in teams of five players representing management teams of different insurance companies (see Figure 2). Competition among insurance companies (teams) is aimed at increasing their market value by making appropriate decisions. They have three tries. They all receive a certificate for participating, and the session and competition winners receive material gifts.

Figure 1. Online presentation of BugaMAP game. Source: Fundación MAPFRE website [29].

2. Materials and Methodology

This study uses an empirical, theoretical, qualitative, and quantitative methodology. The methodology has consisted firstly of a bibliographic search and analysis to construct an analytical theoretical framework which presented in the introductory chapter. Next, a quantitative methodology has been applied with statistical analysis of the data from the evaluation survey answered online by students after finishing the experience of the game. This is, the statistical data available to evaluate the company–university collaboration in applying gamification into university education with the outcome of the opinion and attitudes questionnaire to the BugaMAP players of the University of Sevilla, academic year 2018–2019.

A qualitative methodology has also been applied, and specifically, two techniques: the participant observation of the collaborating teachers and the analysis of the content of the BugaMAP game itself (objective evaluation). For the analysis of game content, eight variables or gamma elements were selected, which were considered essential for a game to fulfil the functions of player involvement, content learning, and skills training, with fun. The content analysis was descriptive, and checked if the game contained the variable and described its format.

Figure 2. BugaMAP game is a team competition. Source: Fundación Mapfre [29].

At the University of Seville, the activity was done as a special class day, and the professors were accompanied by MAPFRE monitors. The game experience consisted of 10 sessions, lasting 4 h each, in which a total of 200 students of economics courses from various degree plans of the University of Seville participated (see Table 3). It took place between days 12 and 20 of March 2018. The activity distributed in the following manner: day 12 (professors meeting), day 13 (1 session), day 14 (2 sessions), day 15 (2 sessions), day 16 (1 session), day 19 (2 sessions) and day 20 (2 sessions).

2.1. Methodological Aspects of the Subjective Evaluation

To collect opinion data about the game (subjective evaluation), a structured response questionnaire was sent to the 200 participants through the Google Drive application, the Google Forms tool that allows questionnaires to be made and statistics obtained from the results of the surveys. As far as reliability is concerned, consistency has been evaluated using the Cronbach alpha coefficient which is the coefficient most widely used to estimate reliability in applied research. Its values range from 0 to 1 [30]. In our questionnaire the value of Cronbach alpha coefficient is 0.942 meaning a high reliability of the instrument used.

The analysis of student opinion data about the game (subjective evaluation) was done based on data collected through an online questionnaire, which was answered by 142 of the 200 students who participated in the game. Using a Likert scale assessment of 5 to 1, they were asked about their satisfaction with the game, the learning achieved, and their opinion regarding various characteristics of the game.

From the 142 participants who answered the questionnaire, of whom 69% were women. Most of the students, 52%, belonged to the degree plan in Labor Relations and Human Resources, and the rest came from degrees plans in Tourism, Statistics, Statistics and Mathematics and Labor Relations and Finance. They were mostly second-year students (64%) and 70% were between 18 and 21 years of age (see Table 3).

Table 3. Sociodemographic characteristics of BugaMEP registered gamers questionnaire sample Univ. Sevilla. All students of economic courses in different degrees, year 2017/18.

By Degree Studying	Statistics	Mathematics & Statistics	Labor Relations & Human Resources	Tourism	Labor Relations & Finances
142	18	19	74	16	15
100%	13%	13%	52%	11%	11%
By Years	First Year	Second Year	Third Year	Fourth Year	Fifth Year
142	16	91	10	22	3
100%	11%	64%	7%	15%	2%
By Age	18/19 Years	20/21 Years	22/23 Years	Other	
142	45	54	22	21	
100%	32%	38%	15%	15%	
By Gender	Women	Men			
142	98	44			
100%	69%	31%			

Source: Elaborated from questionnaire to participants.

2.2. Variables and Indicators for Objective Evaluation

The analysis of qualitative data on the gamification patterns of the BugaMAP game (objective evaluation) was done using a test consisting of eight variables defined with qualitative measurement indicators, based on the contributions of several authors including Chatfield (2010) [13], Gee (2010, 2016) [25], Rojo and Dudu (2018) [12], and Cervera (2012) [24]. See introductory chapter of this article for more detailed information regarding authors' contributions supporting the nine elements characterizing gamification we ended up selecting in order to build a qualitative test for objective evaluation of the BugaMAP simulation game.

The eight qualitative variables selected to test the learning gamification contained in the BugaMAP game are listed below. They are the seven variables proposed by Chatfield (2010) [13] plus the variable "narrative and decision making". For each variable, two to four qualitative indicators that describe it are specified. The indicators are taken from the list of principles of video game design of Gee (2016) [25] (see the list of 24 principles in the Appendix A) as well as the list of skills acquired by the player, according to Cervera (2012) [24].

- Narrative and decision making—This variable is about discerning the experiences for problem solving that the game designs (Gee 1). Each game has a story that enlivens the game, a story that makes the game interesting and motivating in itself. Describing the narrative consists of reconstructing the story, the real story that is simulated in the game, and determining in what territory or place it is framed, who are the characters, what is the role of the player or team of players, what is the objective of the game, and what resources are available to reach the objectives. This part trains skills such as "make your own decisions and take the consequences" (Cervera 8) and "be yourself and live your own experience" (Cervera 7).
- Levels of experience that measure progress—This variable is a very relevant part of the gamification process and consists of the game being able to create flow starting with manageable challenge and low stress (Gee 8); allow for early success and the margin to accommodate larger challenges (Gee 7); lower the cost of failure to encourage exploration and innovation (Gee 19). It is part of the motivation aspect, that the player looks competent because he/she is addressing achievable tasks. This part of the gamification would train for the ability to "take risks, dare to risks" (Cervera4) and to "compete in face of achievable challenges" (Cervera 5).
- Multiples short term and long-term objectives—This variable addresses how the game manages the economy of attention (Gee 2); if the game does a good job of combining the mechanics with the contents (Gee 4); if the game starts with simpler mechanics at the beginning and then goes deeper (Gee 6); and if the game orders the problems well, so as to generate ideas (Gee 10). This part of the gamification trains the ability to "have purposes to accomplish objectives" (Cervera 1) and the ability to "think long term" (Cervera 2).
- Gifts the efforts—This variable addresses analyzing which behaviors are most valued by the game, in which direction the person is motivated, and what awards/rewards are given in the sense that the game relies on learning by doing (Gee 11). These data indicate if the game transmits motivation, involvement, persistence, identity (Gee 3) and that the mechanics of the game have a "fair" functioning (Gee 5).
- Respond quickly frequently and clearly—This variable addresses the game's ability to provide data useful for solving problems (Gee 12); to give a lot of feedback useful for progress (Gee 17); to provide data that advise growth and trajectory (Gee 21); provide data to assure learning is integrative and that there is evaluation (Gee 22). The skill related to this part of the gamification would be to "ask all the possible questions" (Cervera 6).
- Element of final uncertainty (epic moments)—This variable addresses whether the game manages to create the expert cycle, practice-domain-challenge (Gee 9); if the activities connect with achievements and strategies (Gee 13); if systemic thinking is encouraged (Gee 24). By systemic thinking it is understood that the game allows the player to see the whole picture and not just individual actions, which helps the player see how the pieces fit together or can fit together.
- Windows of information to assist with decision making and learning from decision impacts—This variable addresses whether the game places meaning of words with actions (Gee 14); if it gives language and information just at the right moment (Gee 15); if the game creates "fish tanks" and "sandboxes" to reflect on (Gee 16). As seen in the framing theory, information given at the right time is assimilated better.

- Other people–collective game or comparing and commenting with others—This is a variable that can be measured in terms of individual play or team play, if there is a pooling, whether the decisions are individual or in teams. And, according to Gee 18 and Gee 20, it also measures whether the teachers give feedback to the game designers, and if the game offers options and opportunities of customization. This part of gamification addresses training related to the ability to collaborate with others in multiplayer games and the ability to learn to play together (Cervera 10).

3. Results

Results have been classified into two parts: results of the student opinion questionnaire evaluating the BugaMAP game (subjective evaluation); and results of the analysis of patterns of gamification of learning contained in the game (objective evaluation).

As a reminder, in the game BugaMAP each student is part of a management team of an insurance company along with four other students. In each session, the management teams representing several different companies compete with each other to better position their company in the stock market. They have three playing opportunities to make decisions and reach their desired goal.

3.1. Results of the Subjective Evaluation by Questionnaire to Players

The game BugaMAP is fun in the opinion of 88% of the students who answered the questionnaire (see Table 4). Finding the game to be fun, however, does not mean that they consider it easy (only 35%). This indicates that the game manages to be intrinsically motivating in a way that the challenge of overcoming obstacles or difficulties is a pleasant challenge.

For more than 90% of the students, the gradient of difficulty and progression of the game is highly valued as is the contributions of information helpful for decision making and learning from decision impacts. More than 80% of the students consider the game to be well organized, that previous concepts and instructions are clear, that it has a good pace and duration.

Table 4. Subjective perception of the game BugaMAP.

		Agree (4–5)	Neuters (3)	Disagree (1–2)	Total
1	It is fun	88%	10%	2%	100%
2	It is easy to play	35%	51%	14%	100%
3	It is well organized	92%	6%	2%	100%
4	It is creative	85%	13%	2%	100%
5	It is useful for learning	83%	13%	4%	100%
6	The concepts and previous instructions are very useful for decision-making	76%	18%	6%	100%
7	Facilitates and provides business and insurance sector knowledge	84%	12%	4%	100%
8	The duration of the activity has been adequate to acquire the objectives that were proposed at the beginning	68%	19%	13%	100%
9	The number of decision-making opportunities is adequate	80%	17%	3%	100%
10	The contents developed during the training activity have been useful and have been adapted to my expectations	71%	25%	4%	100%
11	It reflects the business reality (insurer)	83%	16%	1%	100%
12	The knowledge acquired will be useful in the future	70%	23%	7%	100%
13	Value the acquisition of soft skills (management of your work time, leadership and teamwork)	79%	19%	2%	100%

The game is considered by students to be creative, but it also is valued as a realistic simulation. The majority of students appreciate that the game represents the reality of the current sector and that the knowledge acquired will be useful in the future.

Opinion of the students about skill training has also been evaluated. A total of 79% of the participants agree that the game does a good job of training people for public speaking, teamwork, and the management of decision-making time.

Finally, there are the results from the part of the questionnaire regarding self-perceived satisfaction (see Table 5). A total of 85% of the students indicate overall positive satisfaction with the game. They considered that their expectations and the objectives were met, and 88% indicated that they would recommend participation in a future BugaMAP activity to their colleagues.

Table 5. Self-perceived satisfaction of the BugaMAP game.

		Agree (4–5)	Neuters (3)	Disagree (1–2)	Total
1	Overall satisfaction	85%	14%	1%	100%
2	The expectations I had regarding the usefulness of the BugaMAP simulation game in which I participated have been met	80%	16%	4%	100%
3	In general, I am satisfied with the development of	83%	16%	1%	100%
4	I would recommend this training activity to other colleagues	88%	9%	3%	100%

3.2. Results of the Objective Evaluation of the Gambling Experience

Below are the results of the objective evaluation of the gamification of the game BugaMAP, which is broken down into eight variables.

3.2.1. Narrative and Decision Making

Our player holds a high professional position in an insurance company and has power and autonomy to take well informed decisions inside a board of directors. The context of the BugaMAP simulation game is that of a closed market in which several teams representing different insurance companies compete against each other in one or more branches: automobiles, multi-risk, health, civil liability, and companies. The objective is for each insurance company to maximize the value of its shares in relation to the others owing to the correctness of its business management decisions.

Each team is composed of five players representing the Board of Directors of the company. The positions of CEO, Financial Director, Risk Director, Human Resources Director, and Director of Administration are distributed among the players, and members have to agree to make decisions as a team.

Each insurance company competing in a session starts the game with the same market share, number of customers, insurance prices, payroll, etc., and with a value of 100 points per share. Throughout the game, each team of players will have to make decisions about its pricing policy, underwriting, distribution, remuneration to sellers, distribution of investments, level of expenses and reinsurance [28].

3.2.2. Levels of Experience that Measure Progress

The levels of experience correspond to the four stages of the game throughout its four-hour duration (see Figure 3). The first stage is the introduction and presentation of the game and the economic concepts of the insurance sector. The next three stages all begin with a decision-making assignment, followed by a simulation and common analysis of the results. The game ends with a final presentation of results, a colloquium, and conclusions.

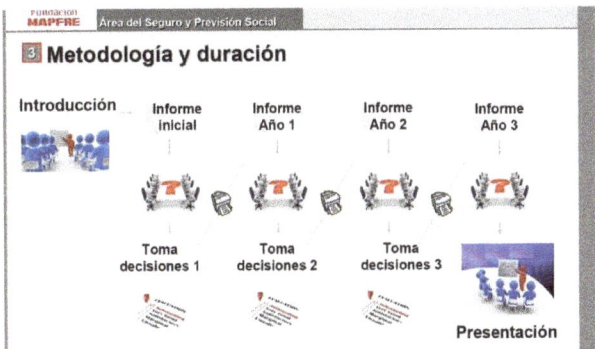

Figure 3. The four stages of the BugaMap game. Source: Fundación MAPFRE [29].

Each team of players can see how its insurance company performed after a first-year of exercise and how it compared with other companies that made different decisions. In addition, in the analysis of the results of each stage, a representative of each team explains why the decisions were made and analyzes the results and their consequences.

As the game continues for two more exercises, each team of players will see their progress and the intermediate results they are achieving. In addition to the decisions as a team, each insurance company will be affected in the game by the uncertainty of the markets and factors exogenous to the company that can combine against or in favor of its annual results.

3.2.3. Multiple Short-Term and Long-Term Objectives

In the short term, the objectives are to become familiar with the economic concepts and content of the activity, how decisions are made in insurance companies, and the competitive market in which companies operate. That is, to develop an understanding of complexity.

And, in view of the impact of their decisions on the company's overall results and the causes of their intermediate results, each business team gets involved in understanding the impact that some decisions may have when combined with others and external risks that can affect the insurance companies' profitability. In this way, teams advance towards the objective of achieving the ultimate success of maintaining or increasing the value of their company via their actions and decisions.

The business team that achieved the best results in the BugaMAP tournament at the University of Seville was the team known as "Europe", which had an ending share value of 97.5 points. This emphasizes the difficulty of the game given that each team started with a price of 100 points per share. And the simulated insurance company that won the BugaMAP competition at the University of Seville was composed of five second-year students.

This was a surprise, because the MAPFRE Foundation considered the game to be aimed at professionals, or students in their final year of a Master's degree program. The fact that students with a lesser educational base and knowledge of economics managed to win the tournament may be explained by the overall vision of gamified learning which is offered in a way that facilitates complex intuitive thinking that speeds up decision-making. Conversely, those students with formative structures of information accumulation may see their decision-making slowed down.

3.2.4. Gifts for Efforts

The prize structure of the BugaMAP game shows that it mainly relies on its own narrative and fair functioning. The game, as it is aimed at professionals in the insurance sector or students in the area of economics, and the prize structure acts as an extrinsic motivation to attract players to get involved in. For simply finishing the game, the players were rewarded with a certificate of participation issued by the MAPFRE Foundation. For winning the game within a session, each member of the winning team received a pen drive. For winning the tournament each member of the winning team was awarded 200 Euros in Amazon shopping cards.

3.2.5. Respond Quickly, Frequently, and Clearly

The BugaMAP game is based on learning by doing, provides data for the resolution of decision-making problems faced by players and provides data that inform the growth and trajectory of playing. But, the BugaMAP game system of providing that information is probably not as fast or as often as is found in commercial video games. The BugaMAP game uses a way of providing information that allows for discussion within all the teams that participate in a game session.

There are several particularly relevant moments when the game provides data. Data are provided at the beginning of the game during the briefing in which, in addition to explaining the concepts and narrative of the game, players are given brief written explanatory material. At the beginning of the first annual exercise each team is given strategic and economic documentation on the insurance market. Decisions of each team are based on the analysis of this documentation.

Later, the decisions made by each business team are acted on by the game software to produce results that constitute a response of the game and a contribution of information on the impact of the team's decisions on the company's profit and loss account. That information on results is provided by the game in each of the three waves (annual exercises) of the game.

Finally, the game generates and provides data on serious claims that happen randomly and on the occurrence of a catastrophic event that causes a rise in the claims rate thus impacting the company's income statements.

3.2.6. Element of Final Uncertainty (Epic Moments)

The game has its great epic moment at the end, because that is when all results from all decisions made by each management team are resolved and hierarchy of ranking is determined. But, there are also epic moments of lesser intensity at the end of the two previous waves, because each team can see the impact of their decisions. Epic moments can also be considered those that are experienced within each business management team when making decisions. And, there is no doubt that the news of catastrophes or external events provided by the game are also times of impact due to the uncertainty about the results.

The BugaMAP game is full of emotion-provoking moments. As decisions are made, the teams are practicing and learning how to play the game better, although there is still uncertainty until the last match as to what a team's final position will be. If they adopt their strategies properly, they can achieve the expected results. Throughout the different decision-making processes, they can modify their strategies if they consider that those previously adopted have not been the right ones. There is feedback among the players, and the results of decisions made in one phase of the game allow them to learn and better define future strategies, until the great epic final moment arrives.

3.2.7. Featured Attention Windows. Information Contributions

The BugaMAP game uses written information instead of attention-grabbing windows built into the software. It has already created those elements of language and information it has to provide at any given moment, but it still needs to develop the software so that players can access the information virtually.

In the game version of this experience, Fundación MAPFRE's monitor played the role of providing information and resolving doubts at the precise moment. It was also necessary for the monitor to enter data of the team's decisions into Fundación MAPFRE's computer. This is something that has already been modified in the new version of the game; the foundation provides a computer to each team to enter the data of their own decisions, thus speeding up the process.

It can therefore be considered that the game BugaMAP is still under development especially with regard to information inputs to and outputs from the game.

3.2.8. Other People–Collective Gamers or Comparing and Commenting with Others

The very nature of the BugaMAP game emphasizes its character as a collective game that makes it possible to train users in the abilities of communication, oral presentation, and team work. Because the game is established as a competition between companies, a player needs four other players to form the management team of a company. This means that continuously, throughout the game, players are debating and commenting on the problems they face and the possible decisions to be made.

After each of the three waves of the game there is a pooling in which each company presents to the other four companies involved in the game, how they arrived at the results of their company and why they made their decisions. Hence, the collective exchange expands from 5 to 25 people interacting.

Also, the university professors and the MAPFRE Foundation monitors take part in the results debates and interact with the players to clarify doubts and provide information that feeds back to the course of the game.

All of this interaction and communication is focused mainly on the objectives of the game, but also addresses aspects of how to improve the game itself in its digital format. Therefore, it can be considered that both the professors of the University of Seville and the students who were involved in the sessions, contribute to the MAPFRE Foundation as testers of the game and collaborators in its further development.

In all practicality, personalization that the game allows is currently limited to choosing the insurance branches preferred by each business team.

4. Discussion

This study focuses on gamification opportunities for university teaching with specific attention to the option of company-university collaboration for implementing and developing serious videogames or gamified experiences. In addition to reviewing a selection of literature on the state of the art regarding the use of gamification for education, this study has addressed the case of collaboration between Fundación MAPFRE and University of Seville in applying the Foundation's game BugaMAP to university students training about insurance.

With the support of several authors, we have shown that the advances in gamification that have revolutionized and made a world-wide $138 billion dollar video game industry, with 8% annual growth and 2.3 billion players, consist fundamentally of a set of techniques that succeed in involving people in complex learning and skill-acquisition activities [2–4,7,13]. Outside the entertainment industry, some economic activity sectors such as education have become interested in the skills acquisition aspects of gamification [2,26], while others such as marketing and services have become more interested in the fidelity or engagement of the users [5].

The diversity of interests and impacts that surround gamification leads us to consider that it, as a technological advance, represents a great new paradigm in the history of knowledge acquisition. Transversality of the scientific disciplines involved is wide, both because of new multidisciplinary scientific knowledge and because of economic and productive performance. The fields of knowledge that combine gamification range from computer science and mathematics, through neurology and graphic art, to behavioral sciences, marketing, sociology, pedagogy and communication [1,4,6,14,15].

In recent years, game technology has been in a process of transference, in the form of serious games or gamified experiences, towards many other sectors such as defence and security, health and fitness, commerce, marketing and services, education and training, social welfare utilities, etc. [5,10,11]. The market for serious games isestimated to reach $5.5 billion in 2020 [11]. Currently, only one serious game is produced for each five produced for entertainment; however, game industry reports for 2018 show that there has been a significant increase in demand for gamification in all sectors [3,7].

Regarding institutions involved in producing serious video games, data from research done on free, serious environmentally-oriented videogames indicate that large companies, major NGOs, international organizations and universities tend to be leaders in serious games production in the North American market [8]. Spain is similar except for the involvement of universities. We find that at present universities in Spain are seldom connected with the production of serious video games and that public administrations avoid budgetary commitments to support this involvement [9,12]. Those free online access videogames on environmental issues analyzed by Ouariachi 2017 [9] were tested with students in research done by Rojo & Dudu 2018 [16] using the Gee 24 criteria list. University students from third year questioned their quality and found them far from attractive or interesting. The majority of serious video games were criticized by students on the basis of their narrative, goals, decision-making opportunities for the player, simplicity, the impossibility of recovering from bad results, etc. [12,16].

This seems to be different with regard to companies that strive to develop a quality game without giving free or commercial access to other users. The experience analyzed by the MAPFRE Foundation with the BugaMAP game shows that most companies are considering their game a patrimonial investment destined for the training of their own technicians. The opinion of the Seville students who played theBugaMAP, however, was very different. For more than 80% of the students the game was fun, organized, creative, not easy, useful, provided them with knowledge reflecting the business reality of the insurance sector, had an adequate number of decision-making opportunities, met their expectations and they would recommend it to other students [31]. But, such a product as BugaMAP may have taken the company more than a decade to develop, and since the collaboration with the University started Fundación MAPFRE has introduced improvements every year. That is why collaboration with universities was seen by the MAPFRE company as part of an improvement strategy and also as a way to disseminate knowledge about an industry, such as the insurance industry in this case.

The formula of collaboration between universities and companies in the development of video games owned by the latter has proved to be of interest for both companies and universities.

For universities, the interest lies in the fact that this collaboration enables them to introduce educational innovations and experiential learning methods into the classrooms. An additional interest is that these collaborations are formative for professors, because teachers in general and university professors in particular find few opportunities to train in digitalization and gamification of content. Other authors subscribe to the idea that "teachers or professors and game designers need to work together" [2,25].

Professors collaborating with companies see the video game industry as an ally to gamify and spread with fun the teaching skills in their respective subjects that have taken them so many years to acquire. As the results of the BugaMAP case study show, contrary to the notion that the use of gamification in education would lead to replacement of the teacher, the introduction of a serious game in university learning may require collaboration among several teachers and the help of professional technicians (monitors) familiar with the rules, process and software of the game. This means that at least in the stages of development of gamified products or serious video games for the university, there would be an increase in the demand for specialists rather than a decrease.

In reference to the benefits for companies collaborating with universities for the development of video games or gamification, the BugaMAP case study presented here showed that the company that owns the game was able to identify areas of potential improvement for developing a more advanced version of the game for the following year. Thus, this type of collaboration is also beneficial for the company. Among the weaknesses of the game that were identified, the most important were in the domain of human-computer interaction, time delay in scoring and that the information necessary for advancement of the game was delivered on paper and explained verbally. As a result of this collaboration, the version that would be used in the University of Seville during the following academic year presented a more advanced virtual format.

Concerning the way gamification works, there is general agreement on the key elements that reward the brain and engage the participant through forms of intrinsic motivation (e.g., narrative, autonomy, decisions, goals, framed information, relations) and extrinsic motivation (e.g., gifts, batches, rewards) [13–15,18–20]. Quality gamification (design of video games, serious games or gamified experiences) has high costs, but its effectiveness rests on its quality. Gamification is seen as capable of reducing significantly the time required to train a person with new professional skills or behaviors by creating virtual environments that simulate real experiences [27]. In the case of the BugaMAP game, for example, four hours were enough for a team of five economics students to experience the equivalent of one year of behavior as well informed responsible executives who make complex decisions about the management of an insurance company [28].

In this study, eight fundamental elements were identified as basic content for an acceptable level of gamification in game design. They were selected on the basis of the contributions of Chatfield 2010 [13], Gee 2016 [25] and Cervera 2012 [24], which compared well with the considerations of other authors [2–4,6,10,16–22]. This list proved to be a valuable tool for evaluating qualitative data from the BugaMAP case study, and it also validated the contributions of the authors above mentioned. The list of variables is as follows: narrative and decision-making; levels of experience that measure progress; multiple short-term and long-term objectives; gifts for efforts; respond quickly, frequently and clearly; element of final uncertainty (epic moments); windows for information to assist with decision-making and learning from decision impacts; and other people (playing with others and commenting with others).

Results of the objective evaluation of the BugaMAP game show that some elements of its gamification or game technology are more developed than others. Among its strongest points are narrative and decision making as well as commenting on or relating to others. Its weakest points arethe speed of the response, the gifts for efforts and the information windows. Other issues that this test reveals are that the part related to the interaction with the HCI computer and the graphic design of the BugaMAP game left out of the evaluation would require a specific study in itself [4,25].

For future research, it is necessary to go further in finding specific indicators that measure the differences in level of advancement in gamification relative to each of the variables or elements of gamification identified in this study. This is important for comparing differences among games in a more precise way, and is also necessary for establishing a way to measure the aspects of graphic design and human-computer interaction elements of a game. Another research issue that arises from this study is to inquire about the factors that influence the slowness with which the educational system is incorporating innovations in gamification. In particular, the values and attitudes of public opinion towards videogames in education and public administration discourse are poorly studied.

An educational video game or simulation of learning experience requires a significant investment of time and knowledge from initial product conception to design, computer programming and essay production. On the other side, only those learning instruments based on well-designed games or games of relative quality are usually satisfactory for students who are products of a digital culture. Thus, if public university institutions were to develop their own videogames, they would require the support of public administrations and companies to cover the high financial cost. The case studied here shows that in Spain and similar serious games markets, a feasible option at present for university professors and students to become familiar with content gamification techniques is through collaboration with those large companies that produce and/or have serious games for education and professional skills training. Likewise, such collaboration could extend to game design enterprises.

Author Contributions: Conceptualization, T.R. and M.G.-L.; Data curation, M.G.-L. and A.R.-R.; Formal analysis, T.R. and A.R.-R.; Methodology, T.R. and M.G.-L.; Supervision, M.G.-L.; Validation, A.R.-R.; Writing–original draft, T.R.; Writing–review & editing, T.R.

Funding: Fundación MAPFRE had no role in the design of the study; in the collection, analyses, or interpretation of data; in the writing of the manuscript, or in the decision to publish the results.

Acknowledgments: We gratefully acknowledge the support provided by Fundación MAPFRE in making possible the application of the game BugaMEP at the University of Sevilla. Also, we acknowledgments to the University of Sevilla and Fac. Ciencias del Trabajo for providing the space for the competition and emailing service for recruiting students to participate in the seminars as an extracurricular activity.

Conflicts of Interest: The authors declare no conflict of interest.

Appendix A 24 Principles of Video Game Design for Learning (Gee 2016, UNESCO)

1. Design experience for problem solving
2. Manage the economy of attention
3. Motivation, involvement, persistence and identity
4. Game mechanics more content, well combined
5. "Fair" operation of game mechanics
6. Simpler mechanics at the beginning and deepening
7. Early success and margin to accommodate larger challenges
8. Create flow. Start with manageable challenge. Little stress
9. Create the expert cycle: practice-domain-challenge
10. Order problems well: generate ideas
11. Relying on learning by doing
12. Give data, for problem solving
13. That the activities connect with achievements and strategies
14. Place meaning of words with actions, etc.
15. Give language and information just at the right time
16. Create "fish tanks" and "sand boxes" to reflect on
17. Give lots of feedback on your progress
18. Teachers give feedback to game designers
19. Lower the cost of failure to encourage exploration and innovation
20. Offer options and customization
21. Advise growth and trajectories with data
22. Integrative learning and evaluation (with data)
23. Stimulate the modification and doing (be them designers)
24. Encourage systemic thinking

References

1. Hamari, J.; Koivisto, J.; Sarsa, H. Does gamification work? A literature review of empirical studies on gamification. In Proceedings of the 47th Hawaii International conference on System Science, Hilton Waikoloa, HI, USA, 6–9 January 2014; IEEE Computer Society: Washington, DC, USA, 2014. [CrossRef]
2. Nah, F.; Telaprolu, V.; Rallapalli, S.; Venkata, P. Gamification of Education using Computer Games. In *HCI 2013, Part III. LNCS*; Yamamoto, S., Ed.; Springer: Berlin/Heidelberg, Germany, 2013; Volume 8018, pp. 99–107.
3. Newzoo Consultancy. *Global Games Report 2018*; Short Free Version; Newzoo Europe: Amsterdam, The Netherlands, 2019.
4. Deterding, S.; Dixon, D.; Khaled, R.; Nacke, L. From game design elements to gamefulness: Defining gamification. In Proceedings of the 15th International Academic MindTrek Conference: Envisioning Future Media Environments, Tampere, Finland, 6–8 October 2010; ACM: New York, NY, USA, 2011; pp. 9–15.
5. Hamari, J. Do badges increase user activity? A field experiment on the effects of gamification. *Comput. Hum. Behav.* **2017**, *71*, 469–478. [CrossRef]
6. Dicheva, D.; Dichev, C.; Agre, G.; Agelova, G. Gamification in Education. A systematic Mapping Study. *Educ. Technol. Soc.* **2015**, *18*, 75–86.
7. Desarrollo Español de Videojuegos. *Libro Blanco del Desarrollo Español de Videojuegos 2017*; Desarrollo Español de Videojuegos (DEV): Madrid, Spain, 2018.
8. Katsaliaki, K.; Muysafee, N. Edutainement for sustainable development: a survey of games in the field. *Simul. Gaming* **2015**, *46*, 647–672. [CrossRef]

9. Ouariachi, T.; Olvera-Lobo, M.D.; Gutiérrez-Pérez, J. Gaming Climate Change: Assessing Online Climate Change Games Targeting Youth Produced in Spanish. *Procedia Soc. Behav. Sci.* **2017**, *237*, 1053–1060. [CrossRef]
10. Michaud, L.; Alvarez, J.; Alvarez, V.; Djaouti, D. *Serious Games: Training & Teaching–Healthcare–Defence & Security–Information & Communication*; IDATE: Montpellier, France, 2010.
11. Markets and Markets. Serious Game Market Worth $5448.82 Million by 2020. Recuperado el 25 de Febrero de 2017, de Markets and Markets. 2015. Available online: http://www.marketsandmarkets.com/PressReleases/serious-game.asp (accessed on 8 September 2019).
12. Rojo, T.; Dudu, S. Los videojuegos en la implementación de políticas de mitigación del cambio climático. *Ambitos Revista Internacional de Comunicación* **2017**, *37*, 1–25.
13. Chatfield, T. *Fun INC: Why Games Are the 21st Century's More Serious Business*; Virgin Publishing Ltd.: London, UK, 2010.
14. Grimaud, E. *Beau-Bien-Bon: La Formule Magique Pour Sourire à la vie!* Marabout: Marabout, France, 2017.
15. After Watching this Your Brain will not be the Same. Available online: https://www.youtube.com/watch?v=LNHBMFCzznE (accessed on 18 September 2019).
16. Rojo, T.; Dudu, S. Los "juegos serios" como instrumento de empoderamiento y aprendizaje socio-laboral inclusivo. *Revista Fuentes* **2017**, *19*, 95–109. [CrossRef]
17. Lakoff, G. *Don't Think of An Elephant. Know Your Values and Frame the Debate*; Chelsea Green Publishing: White River Junction, VT, USA, 2004.
18. Kapp, K.M.; Blair, L.; Mesch, R. *The Gamification of Learning and Instruction Fieldbook Ideas into Practice*; Wiley: San Francisco, CA, USA, 2014.
19. Ryan, R.M.; Deci, E.L. Intrinsic and Extrinsic Motivations: Classic Definitions and New Directions. *Contemp. Educ. Psychol.* **2000**, *25*, 54–67. [CrossRef] [PubMed]
20. Werbach, K.; Hunter, D. *How Game Thinking Can Revolutionize Your Business*; Wharton Digital Press: Philadelphia, PA, USA, 2012.
21. Frasca, G. Los Videojuegos Enseñan Mejor que la Escuela. Tedxtalks Junio de 2012 Monteviedo. 2012. Available online: https://www.youtube.com/watch?v=TbTm1Lkm18o&t=11s (accessed on 8 September 2019).
22. Smethurst, T.; Craps, S. Playing with Traume: Interreactivity, Empathy and Complicity with the Walking Dead Video Game. *Game Cult.* **2015**, *10*, 269–290. [CrossRef]
23. Nah, F.-H.; Zeng, Q.; Telaprolu, V.R.; Ayyappa, A.P.; Eschenbrenner, B. Gamification of Education: A Review of Literature. In *HCI in Business*; LNCS, Nah, F.-H., Eds.; Springer: Berlin/Heidelberg, Germany, 2014; pp. 401–409.
24. Cervera, H. Diez Cosas que Aprendí de los Videojuegos. Enero de 2012 [Vídeo de Youtube] Recuperado de. 2012. Available online: https://www.youtube.com/watch?v=Q4nFUFO_rXw (accessed on 8 September 2019).
25. Design Principles for Video Games as Learning Engines. Available online: http://mgiep.unesco.org/wpcontent/uploads/2016/02/3-webinar-reading-reference.pdf (accessed on 8 September 2019).
26. Contreras, R.; Eguia, J. (Eds.) Gamificación en las aulas universitarias; (Bellaterra). Barcelona: Universidad Autónoma de Barcelona. Recuperado a partir de. 2016. Available online: http://incom.uab.cat/download/eBook_incomuab_gamificacion.pdf (accessed on 8 September 2019).
27. McGonigal, J. *SuperBetter: A Revolutionary Approach to Getting Stronger, Happier, Braver and More Resilient—Powered by the Science of Games*; Penguin: New York, NY, USA, 2015.
28. González-Limón, M.; Serrano-García, A. Innovación en las Enseñanzas Universitarias Para el Aprendizaje del Sector Empresarial Asegurador: Una Experiencia con BugaMAP en la Universidad de Sevilla. In *Innovación en la Práctica Educative*; Egregius Ediciones: Sevilla, Spain, 2018; pp. 212–226.
29. Fundación Mapfre Website. Available online: https://www.fundacionmapfre.org/fundacion/es_es/programas/formacion/congresos-jornadas/juego-estrategia-seguros-bugamap.jsp (accessed on 8 September 2019).
30. Cronbach, L.J.; Shavelson, R.J. My current thoughts on coefficient alpha and successor procedures. *Educ. Psychol. Meas.* **2004**, *64*, 391–418. [CrossRef]
31. Encuesta BugaMAP 2018 de opinión y actitudes de los estudiantes de la Universidad de Sevilla sobre BugaMAP. 2018. Available online: https://docs.google.com/forms/d/1EPZen9vYYvRRRqMzMQsY-Z-9yPk3lOoAS5uIPYg9co8/viewform?edit_requested=true (accessed on 18 September 2019).

© 2019 by the authors. Licensee MDPI, Basel, Switzerland. This article is an open access article distributed under the terms and conditions of the Creative Commons Attribution (CC BY) license (http://creativecommons.org/licenses/by/4.0/).

Article

Usability and Engagement Study for a Serious Virtual Reality Game of Lunar Exploration Missions

Lizhou Cao [1,*], Chao Peng [1] and Jeffrey T. Hansberger [2]

1 School of Interactive Games and Media, Rochester Institute of Technology, Rochester, NY 14623, USA; cxpigm@rit.edu
2 Army Research Laboratory, Huntsville, AL 35899, USA; jeffrey.t.hansberger.civ@mail.mil
* Correspondence: lc1248@rit.edu

Received: 29 June 2019; Accepted: 29 September 2019; Published: 3 October 2019

Abstract: Virtual reality (VR) technologies have opened new possibilities for creating engaging educational games. This paper presents a serious VR game that immerses players into the activities of lunar exploration missions in a virtual environment. We designed and implemented the VR game with the goal of increasing players' interest in space science. The game motivates players to learn more about historical facts of space missions that astronauts performed on the Moon in the 1970s. We studied usability and engagement of the game through user experience in both VR and non-VR versions of the game. The experimental results show that the VR version improved their engagement and enhanced the interest of players in learning more about the events of lunar exploration.

Keywords: serious game; usability; game engagement; virtual reality

1. Introduction

Since the advent of head-mounted displays (HMDs), virtual reality (VR) technologies have been used widely in digital media and entertainment. The low cost of HMDs makes VR accessible to the public. Many VR games using HMDs and handheld motion controllers have been released to the public in recent years. Games developed with VR technologies can provide a highly immersive experience [1]. Serious games are "(digital) games used for purposes other than mere entertainment" [2]. They are used in many areas such as education, healthcare, ecology, and scientific research [3]. Adopting VR technologies in serious games can simulate a learning or training environment that would otherwise be impossible for people to access in the physical world.

In this work, we design and implement a serious VR game that immerses players into activities of lunar exploration missions in a virtual environment. The game is based on the historical events of the Apollo 16 mission in the 1970s, in which astronauts landed on the Moon, drove the lunar rover, and eventually returned to the Earth. In this game, we implement planning, preparing, and driving activities of the Apollo 16 mission by utilizing VR technologies and handheld motion controllers. In our previous work, we developed a similar serious game [4], but it was a non-VR version playable with a single-screen display and keyboard–mouse/joystick inputs. Besides the contributions in game design and development, we investigate how the VR technology effects usability and game engagement compared to the non-VR version. In our study, we utilized a between-subject design, where half of the participants play the non-VR version of the game and the other half of the participants play the VR version of the game. We used the Game Engagement Questionnaire (GEQ) [5] and an interview questionnaire to measure their levels of engagement. The results show that the VR version of the lunar roving game took longer for participants to finish but enhanced the game engagement and their motivation to learn the events of lunar exploration.

The paper is organized as follows. Section 2 describes existing work on VR-based serious games. Section 3 describes the background of lunar exploration missions by The National Aeronautics and Space Administration (NASA) in the United States in the late 1960s and early 1970s and their influence on education. Section 4 discusses the design details and gameplay mechanics of the VR game. Section 5 describes the design of the usability and engagement study. Section 6 presents the analysis results, and we conclude the work in Section 7.

2. Existing Work in Serious VR Games

Our Lunar Exploration Game is a serious VR game with the goal of motivating players to learn historical events of the Apollo 16 mission. Serious games have been applied in many domains such as education, healthcare, training, and scientific exploration. In Section 2.1, we review papers that discuss serious games used in different domains. In Section 2.2, we review studies evaluating the effectiveness of serious VR games, with a focus on engagement, usability, and learning outcomes. In Section 2.3, we review the theories underpinning usability and engagement studies in serious VR games, which have impacted the design of our experiment.

2.1. Applied Domains of Serious VR Games

Susi et al. [2] gave a review of serious games. They presented the definition, domain, history, and market status of serious games. Djaouti et al. [3] classified serious games with the Gameplay/Purpose/Scope (G/P/S) model to combine both "serious" and "game". Mikropoulos and Natsis [6] gave a review of educational virtual environments from the year 1999 to 2009. They identified several features in VR for learning and discussed reasons why VR could be useful in a learning situation. The most important reason they identified is that VR provides the possibility for users to try situations not accessible or dangerous in the real world. Potkonjak et al. [7] presented a review of virtual laboratories for science, technology, and engineering education. They argued that virtual labs can offer advantages that real labs may not provide. The advantages included cost-efficiency, flexibility, multi-user support, and the access of dangerous or not accessible situations. Alhalabi [8] discussed the effectiveness of VR systems for engineering education. They evaluated traditional education approaches, Corner Cave systems, and HMD systems. They found that Corner Cave and HMD systems were better than traditional education approaches in terms of improving students' performance, and the effectiveness of HMD systems was more significant.

However, researchers have shown that in different situations, VR technology does not always play a positive role. Jensen and Konradsen [9] presented a review of the use of HMDs in education and training. They found that using an HMD led to a better result only in a small number of cases. Beyond those cases, using an HMD had no advantage or even performed worse than the non-VR setting. Freina and Ott [10] reviewed the use of immersive virtual environments in education. They discussed the advantages and drawbacks of using VR in children's education and rehabilitation for cognitive disabilities. Virvou and Katsinois [11] discussed the usability of using VR games for education in the classroom. They argued that learning outcomes are different among students since they usually have different game-playing backgrounds, but in most cases, the differences in their backgrounds do not influence their motivation.

2.2. Use Case Studies of Serious VR Games

Mei and Sheng [12] presented a system for virtual hospital-situated learning. Their system was primarily used for learning human organ anatomies. They obtained the result that the situated learning can improve users' motivation. Cheng et al. [13] presented a VR game to teach language and culture. The result showed that the VR game could increase user engagement, but the result did not show a significant improvement in the language learning outcome when the VR game was used. Adamo-Villani and Wilbur [14] presented a study of the virtual learning environment for deaf and hard-of-hearing children. They compared a immersive version and non-immersive version; however, the results of the study did not show a significant difference between the versions. Parmar et al. [15] compared the HMD-based metaphor viewer with the desktop-based display for training users in electrical circuitry. Their results showed a significant learning benefit when using the HMD. They argued that participants may gain better performance and enjoy using the HMD version, but they did not provide an evaluation to support this argument. Greenwald et al. [16] presented a usability study to compare the learning in VR with the learning on a traditional 2D screen. The comparison result did not show a significant difference in quantitative learning measurements, though the completion time of the VR version was always longer than the non-VR version. Olmos et al. [17] provided an educational platform to study the influence of emotional induction and level of immersion on learning motivation. They presented a usability study between an immersive version with an HMD and a low-immersive version with a tablet. The results indicated that the immersive condition and the positive emotional induction can influence knowledge retention. Zizza et al. [18] designed and implemented a multimodule VR learning environment over the network. They evaluated user experience in the VR version by comparing it to a desktop version. Although they stated that the feedback on the VR version was positive, the interview questions used to obtain feedback should be more comprehensive. Buttussi and Chittaro [19] conducted a two-week study on three different types of display settings for learning, including a desktop VR platform, an HMD with a narrow field view and a 3-DOF tracker, and an HMD with a wide field of view and a 6-DOF tracker. They found that HMDs can increase user engagement. In their study, HMD did not lead to a significant increase in knowledge and self-efficacy.

In summary, the use of VR technologies for serious games is a new area. In different use cases, people may have different experiences and opinions about VR. Thus, there is no consensus. Our literature search did not find any studies about the adoption of VR technologies for learning historical events of lunar exploration or other events in space science.

2.3. Usability and Engagement Studies in Serious VR Games

Bowman et al. [20] discussed several evaluation methods according to three key characteristics: involvement of representative users, context of evaluation, and types of results produced. The results showed that the evaluation should be close to the changes between the VR system and non-VR system. Based on their findings, the parameters for usability evaluation that we recorded in our experiment focus on the differences between the VR and non-VR versions of the game. Sutcliffe and Kaur [21] studied the walkthrough method for evaluating VR systems. Key points of this method are a goal-oriented task, exploration and navigation, and system initiative. In our experiment, tasks that the participants complete were based on the walkthrough method.

When measuring the usability differences between a VR system and non-VR system, researchers have employed the between-subjects study design [22,23]. Mcmahan et al. [22] expressed that the reason is to increase experimental controls and reduce confounding variables. In our experiment, we used a between-subjects study design so that each participant was only subjected to the play experience of a single version.

Game engagement is one of the factors that affect learning outcomes and the motivation to play games [24]. Brockmyer et al. [5] developed the Game Engagement Questionnaire (GEQ) to measure the engagement level of players playing a video game. The GEQ has been used in many serious game studies [25–28]. Researchers have also used the GEQ to study VR games [28–30]. In our experiment, we used the GEQ to measure and analyze the engagement level of participants in both VR and non-VR versions of the lunar roving game.

3. Background of NASA-Inspired Serious Games

In the 1960s, students were inspired by the success of lunar exploration missions performed by NASA astronauts and engineers. Photographs and videos of the missions and NASA-developed technologies motivated students to pursue science, technology, engineering, or math (STEM) degrees.

Through the 1960s, the number of students enrolled in science and engineering increased significantly. According to a report by the National Science Foundation [31] and the article by Markovich [32], the percentage of bachelor degrees awarded in science and engineering fields peaked in the late 1960s for the period of 1966–2010.

The experience of learning about lunar exploration missions is usually through a passive setting where information is presented in the form of reading material, speaker talks, videos, museum lectures, or exhibitions. For example, people may explore NASA's website to find a wealth of materials that explain the past and future of lunar exploration. Those materials provide both scientific facts and public narratives for public science engagement. In such a setting, learners passively receive what the assigned material says. In contrast to the passive setting, an active setting gives the learner an opportunity to take a participatory role as he or she is absorbing the knowledge [33]. For example, some summer camp programs promote active learning experiences during a scheduled period of time (e.g., the Space Camp program at the U.S. Space & Rocket Center (Space Camp is located in Huntsville, Alabama, U.S. Website: https://www.spacecamp.com)). However, those summer camp programs usually require a substantial cost for travel or lodging.

In contrast to attending Space Camp, a game of lunar exploration is easier to set up and provides an active learning environment at a lower cost. The complex content of lunar exploration missions can be represented graphically in the game, and the game allows the learner to interact with the content using precisely simulated navigation systems and driving controls based on the operations in real missions. NASA's Eyes [34] is an interactive application developed at NASA's Jet Propulsion Laboratory. It is an educational tool that allows the user to interact with the Earth, solar system, and spacecraft. To maintain a low computational cost on geometry rendering, the Earth and Moon in NASA's Eyes are represented as spheres wrapped with textures, so there is a lack of visual fidelity on geographical features. Another game-based example is NASA Space Place [35], which is a NASA science webpage for kids. NASA Space Place has a few games related to the Moon, rovers, and missions. The games are 2D, cartoonish, and provide only introductory or conceptual information.

NASA supports game developers making NASA-inspired 3D games. NASA has released a variety of 3D models and textures of spacecraft, landing sites, asteroids, etc. [36]. One 3D game is Station Spacewalk Game [37]. It contains a 3D model of the International Space Station that allows the user to experience NASA repair work on the station and learn how the station is assembled. The game runs on a standalone PC or on the web with a Flash plug-in, but it does not work with VR devices. NASA also provides a web-based interactive tool [38] to show the historical landing sites on the Moon and explains briefly the activities that astronauts have conducted. It helps the player to understand the history of lunar exploration, but it does not offer participatory experiences for users. A few years ago, the Immersive VR Education company [39] released an Apollo 11 VR game that is a documentary style VR application presenting historical lunar exploration events through a mix of original audio and video. The National Naval Aviation Museum adopted a similar concept to exhibit the Apollo 11 journey using VR technologies [40] along with a physical Apollo command module. Audiences are immersed in activities such as boarding the rocket, experiencing the launch, witnessing travel through space, and landing on the Moon. Peng et al. [4] gamified the Moon mission and created three playing phases. They presented the lunar roving game on a PC platform and employed a standard screen. They performed a usability study with 30 participants and provided a descriptive analysis of the effectiveness of the game for learning about lunar exploring activities. The results showed that the game enhanced user engagement in NASA-inspired events and increased the users' interest in space science. However, their work did not provide a statistical comparison between a VR-based gameplay and a non-VR gameplay.

4. Game Design Methodology

The concept of game design in this work is similar to the lunar roving game presented by Peng et al. [4]. In this work, we develop a VR version of the lunar roving game, which is compared to the non-VR version presented by Peng et al. [4] in terms of usability and game engagement. The game is composed of three playing phases: planning, preparing, and driving. In the planning phase, the player is guided to create the driving route by placing markers on a virtual lunar terrain map. In the preparing phase, the player selects a subset of devices from the inventory and then loads them onto the lunar rover for later use on the route. In the driving phase, the player drives the rover to the end of the route. During the driving, the player operates a navigation system to determine the driving direction, control the rover's speed, and avoid overheating. The game presented by Peng et al. [4] is a non-VR video game running on a PC with a single screen. The player uses a keyboard/mouse to place markers and select scientific instrument and uses a joystick to drive the rover. In this work, we convert the non-VR version of the game into a VR game. The VR game runs on a PC with the content rendered to a head-mounted display so that the player will feel immersed in a virtual lunar environment. The player plays the VR game using handheld motion controllers.

In general, VR games have many things in common with non-VR video games. For example, both non-VR and VR games produce interactive experiences and require real-time rendering, and their development process is the same. In contrast to the use of a keyboard and mouse in non-VR games, the use of VR head trackers and handheld controllers enhances the freedom of movement for in-game actions. In this section, we discuss the design differences between the non-VR and VR versions of the lunar roving game, the ways of relating to the environment and interface, input modalities for gameplay, and engagement levels in each playing phase.

4.1. Planning Phase

The lunar roving game requires the player to explore a region of lunar terrain on a planned driving route. In the planning phase, the player is asked to identify station stops on the lunar terrain map based on the longitude and latitude values. Both the non-VR and VR versions of the game allow the player to place markers/tokens on the map.

In the non-VR version of the game, the player places markers on the map through a 2D user interface. As shown in Figure 1a, the interface contains a cartoon-style commander who instructs the player how to use the interface. The commander's instructions and longitude and latitude values are displayed as rolling texts in a dialogue box. The player places a marker by mouse-clicking on the map. If the player does not place the marker at the correct location, he or she will be asked to repeat the placement operation for the same marker until it is placed correctly on the map. When all markers are placed correctly, a calculator appears in the interface. The commander will ask the player to calculate the travel time based on the given distance and maximum speed. To do the calculation, the player can enter numbers by mouse-clicking the number buttons on the calculator.

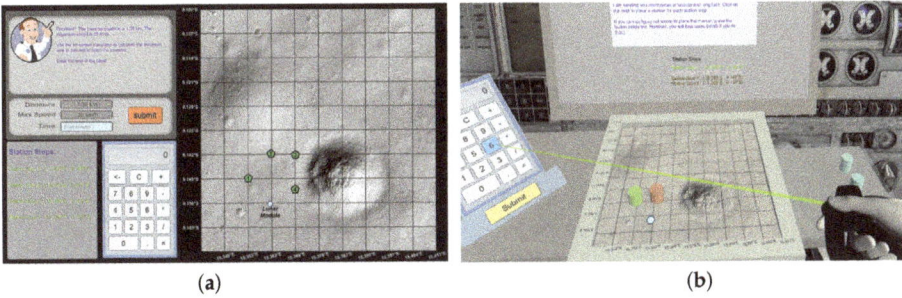

Figure 1. Screenshots of the planning phase. (a) is the user interface for the non-virtual reality (VR) version of the game, and (b) is the 3D environment for the VR version of the game.

In the VR version of the game, the player wears an Oculus Rift headset with which he or she gains a 360 degree interior view of the spacecraft. As shown in Figure 1b, the player can see the lunar terrain map on a desk, and there are four 3D tokens next to the map. The instructions from the commander and the longitude and altitude values are displayed on a virtual monitor in front of the player. In the VR game, the player holds a pair of Oculus Touch controllers in his or her hands, which provide hand presence in the virtual environment. The player can press a button on the Touch controllers to trigger a gameplay event without any interruption from physical hand movements. To place a token on the map, the player points to the token using the controller and then holds a specific button on the controller to pick up the token. Then, the token can be moved across the desk by moving the controller while holding the button. The token can be dropped on the map when the button is released. If the token is not placed correctly, the player is asked to pick it up again and repeat the placement operation until it is placed correctly. When placing a token, the player can lean his or her body forward to have a closer look of the map or lean back to regain the full view of objects on the desk. After all tokens are placed correctly, more instructions from the commander will roll into the virtual monitor; at the same time, a 3D calculator appears on the left side of the desk. To use the calculator, the player points on the number buttons of the calculator using the Touch controller and then triggers a specific button on the controller to press the number button.

4.2. Preparing Phase

In the original Apollo mission, astronauts needed to load communication and scientific instruments onto the lunar rover and use them at station stops on the route. In the preparing phase of this game, the commander describes the related scientific tasks. Based on the description, the player determines what instruments to choose from the inventory and loads them onto the rover. The participant is required to finish a total of five tasks. The first task is to put two astronauts onto the lunar rover. The second task is to load two communication devices. The third task is to load a pallet frame that will hold devices and tools. The fourth task is to load a monitoring device that can be controlled remotely by Houston. The fifth task is to load a camera device to record documentary photos and videos.

As shown in Figure 2a, the non-VR version of the game consists of a 2D user interface. The commander appears in the top-left corner of the screen. The description of related scientific tasks from the commander appears in a dialog box. The player can explore the inventory in the middle panel of the interface. When mouse-clicking on an item in the inventory, the 3D model of the corresponding device will appear in the top part of the right panel. At the same time, the usage of the device will appear in the bottom part of the right panel. When the player checks the checkbox of the item in the inventory, the device will show up on the rover at the location it should be loaded. A 3D view of the lunar rover is shown in the left panel of the interface. The player can rotate the 3D view to look at the rover from different angles.

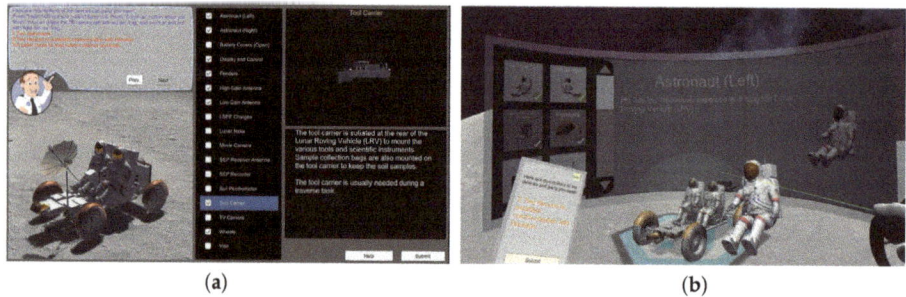

(a) (b)

Figure 2. Screenshots of the preparing phase. (**a**) is the user interface for the non-VR version of the game, and (**b**) is the 3D user interface for the VR version of the game.

The VR version of the game contains a 3D user interface for the preparing phase. Different from the non-VR version, the description of science tasks appears on a virtual display device held in the player's left hand. The player can move the display device closer or farther away from the HMD's viewpoint by moving the Touch controller in the left hand. The device icons and usages are shown in a large curved virtual display. As discussed in Cao et al. [41], a large curved display allows the player to interact with the content in a wide field of view and supports a high level of perceived immersion. The Touch controller in the player's right hand is used to scroll the list of device icons up and down. To view the details of a device, the player points to the device icon and presses a specific button on the controller so that the 3D model of the device and its usage will appear in the right section of the curved display. The player can hold a specific button on the controller to pop the 3D model of the device out of the curved display and move it in 3D space using the controller. The player can drop the device onto the lunar rover, as shown in Figure 2b.

4.3. Driving Phase

In the driving phase, the player drives the lunar rover to explore a geographical region of the Moon. Both the non-VR and VR versions of the game use the same 3D model of the lunar terrain, which is converted from Lunar Reconnaissance Orbiter (LRO) data. The player needs to operate the rover's navigation system to check the driving direction, monitor the speed, and avoid the overheating issue. In particular, the game requires the player to use three navigation devices. The first one is a heading device that displays the driving direction based on the shadow cast on the heading dial. The second one is a speed meter that displays the current driving speed of the rover. The third one is a temperature meter that displays the heat levels of the batteries and motors. The rover has to slow down or stop if the reading of the temperature meter indicates an overheating of batteries or motors.

The interaction with the navigation system is different between the non-VR and VR versions of the game. In the non-VR version, as shown in Figure 3a, a vertical panel is created on the right side of the screen to host the heading device, speed meter, and temperature meter. The 3D model next to this panel is the control console, on which the navigation devices were installed originally. In the VR version, as shown in Figure 3b, the player operates the navigation devices only through the control console on the rover. With the immersive view provided by the HMD, the player can move his or her head closer to the control console to check readings on the devices, just like what astronauts would do when driving the lunar rover.

The real lunar rover uses a T-shaped steering controller for driving. This controller, as a mechanical part of the control console, is located between two seats. It was operated by the astronaut sitting on the left seat. Moving the controller left or right turns the rover to the left or right. Moving it forward increases the speed, and moving it backward decreases the speed. To mimic such special steering control, both versions of the game use a flight stick as the input device, as depicted in Figure 3d.

The camera control in the non-VR and VR versions of the game is implemented differently. In the non-VR version, the default position of the camera is set behind the rover. When the player is driving, the camera will follow the rover's movement. The player can press a button on the flight stick to switch among a few different camera views, such as the side view to see the rover from the side and, the front view. In the VR version, the camera is controlled by the player's head movement, as shown in Figure 3c. The player only has the first-person view of the astronaut sitting on the driver's side. The VR version does not allow view switching.

Both the non-VR and VR versions of the game contain a mini map that shows the location of the Lunar Module (the landing location), locations of the station stops, and geological features of the Moon. In the era of the Apollo missions, there was no GPS system to track where the rover was on the Moon. Therefore, the mini map in the game does not show the location of the lunar rover. However, in the case that the player feels lost, he or she can press a specific button on the flight stick to show the rover's location and its current moving direction only for a few seconds. This can be used a maximum of five times. In the non-VR version of the game, the mini map is at the bottom-left corner, and it is always visible to the player. In the VR version of the game, as shown in Figure 3d, the mini map becomes visible when a button is pressed.

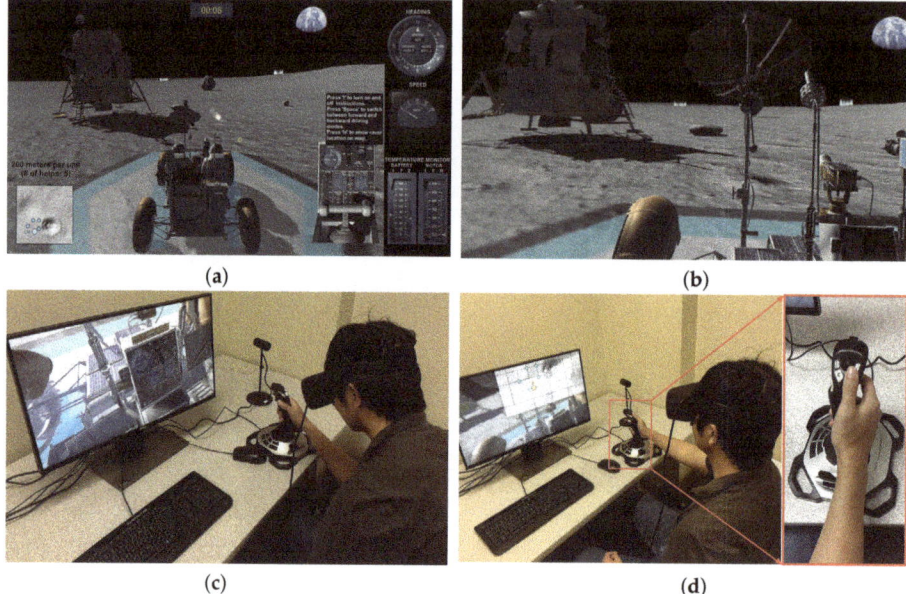

Figure 3. Screenshots and control examples of the driving phase. (**a**) is the 3D environment for the non-VR version of the game, with the navigation system on the right side of the screen, (**b**) is the 3D environment of the first-person view for the VR version of the game, (**c**) shows the head movement with the head-mounted display (HMD) for the readings of the navigation devices in the control console, and (**d**) shows the operation with the flight stick to bring up the mini map in the VR version of the game.

5. Usability and Engagement Study

We designed and performed a study to understand the usability and game engagement of the non-VR and the VR versions of the game. This section describes the design of our study.

We set up the usability and engagement study in a research laboratory in the university. The study took about 30 min for each participant. After arrival, participants first signed the consent form, and then each of them was given a unique user ID. Before playing the game, participants answered a demographic questionnaire about their experience in gaming, virtual reality, space science, and serious games. We utilized a between-subject design, where participants were randomly selected to play the VR and non-VR versions of the game. Each participant was asked to play the same version twice (two trials). After playing the game, they were asked to answer the GEQ [5] (as shown in Table 1) and the interview questionnaire (as shown in Table 2). We recruited a total of 30 participants for the study, 20 males and 10 females. A total of 10 males and 5 females played the non-VR version, and the other 10 males and 5 females played the VR version. Their ages ranged from 20 to 42 years, with the average age of 29.2 years. Twenty-six participants were from STEM majors. Figure 4 shows pictures of participants playing the game during the study. We defined several usability parameters and recorded values of the parameters during the study to evaluate user performance. Table 3 lists the parameters used in the planning, preparing, and driving phases.

Table 1. Game Engagement Questionnaire (GEQ) items [5] and the results of the game engagement evaluation. The rating scale for each GEQ item is 1 (strongly disagree) to 5 (strongly agree).

Questions	Non-VR Version	VR Version
1. I lose track of time.	2.67	3.13
2. Things seem to happen automatically.	2.73	2.93
3. I feel different.	2.60	3.20
4. I feel scared.	1.20	1.48
5. The game feels real.	4.13	4.27
6. If someone talks to me, I don't hear them.	1.80	2.00
7. I get wound up.	1.40	1.93
8. Time seems to kind of stand still or stop.	2.40	2.33
9. I feel spaced out.	2.40	3.53
10. I do not answer when someone talks to me.	1.73	1.93
11. I cannot tell that I am getting tired.	2.27	2.80
12. Playing seems automatic.	3.33	3.13
13. My thoughts go fast.	3.47	3.20
14. I lose track of where I am.	1.80	2.73
15. I play without thinking about how to play.	3.20	3.00
16. Playing makes me feel calm.	3.53	3.40
17. I play longer than I mean to.	2.60	2.87
18. I really get into the game.	4.20	4.27
19. I feel like I just cannot stop playing.	3.53	3.07

Table 2. Interview questions and the results.

	Questions	Rating Scale	Non-VR Version	VR Version
1.	I feel the game is impressive.	1 (strongly disagree)–10 (strongly agree)	8.26	8.60
2.	I feel the game is difficult.	1 (strongly disagree)–10 (strongly agree)	3.13	4.07
3.	After playing this game, do you feel that you want to learn more about historical events of lunar exploration missions?	1 (not at all)–10 (very willing to)	7.67	8.47
4.	After playing this game, do you think you have gained interest in space science?	1 (not at all)–10 (very interested)	7.60	8.33
5.	Do you feel motion-sick during the play?	1 (not at all)–10 (very serious)	2.60	3.90
6.	With respect to the Lunar Module landing location, describe which direction the crater is.	N/A	6 (correct)/15	10 (correct)/15

Table 3. Usability parameters used in our study.

Phases	Parameters	Descriptions
Planning	pl_n	The number of attempts to place all markers or tokens
	pl_{mt}	The total time spent on placing markers or tokens (mm:ss)
	pl_{mt1}	The time spent on placing the first marker or token
	pl_{ct}	The total time spent on calculation (mm:ss)
Preparing	pr_n	The number of attempts to submit
	pr_t	The total time spent in this phase (mm:ss)
	pr_{t1}	The time spent on making the first submission
Driving	dr_t	The total driving time (mm:ss)
	dr_h	The number of times overheating warnings occurred

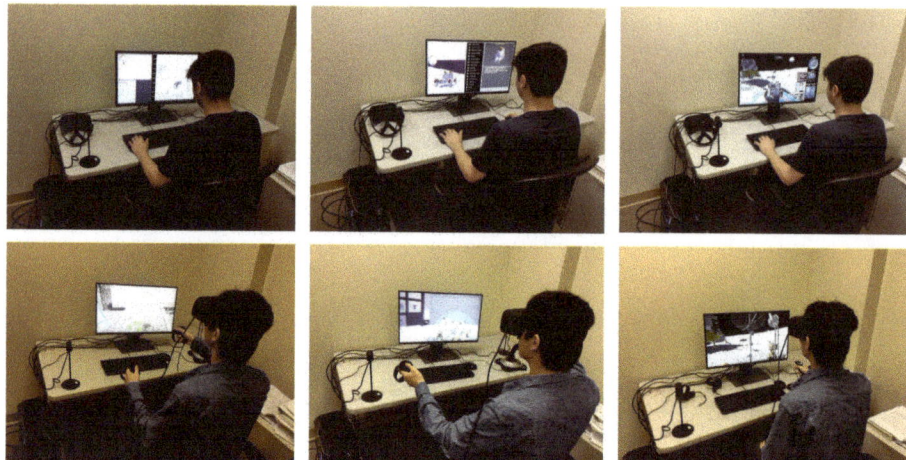

Figure 4. Pictures of participants playing the game during the usability study. The left column shows pictures of participants playing the planning phase. The middle column shows pictures of participants playing the preparing phase. The right column shows pictures of participants playing the driving phase. The first row shows participants playing the non-VR version of the game. The second row shows participants playing the VR version of the game.

6. Evaluation Results

6.1. User Performance

To evaluate user gameplay performance, we applied the analysis of variance (ANOVA) statistical model (with $p = 0.05$) to the values of usability parameters. The format for the time parameter was a tuple of minutes and seconds denoted as mm:ss. Table 4 shows the statistical analysis results between the VR version and non-VR version of the game and the statistical analysis results between the first trial and second trial in each version of the game. The mean values (denoted as M) and standard deviation values (denoted as SD) of the parameters are shown in the "Mean" and "Standard Deviation" columns in Table 4. The "Between-Trial Comparison" and "Between-Version Comparison" columns are the results produced by the ANOVA model. The following subsections discuss analysis details for the planning, preparing, and driving phases between the two versions of the game and between the two trials in each version.

6.1.1. Evaluation of the Planning Phase

We analyzed user performance on two tasks in the planning phase. The first task requires the participant to place 4 markers (in the non-VR version) or 4 tokens (in the VR version) on the lunar terrain map based on given longitude and latitude values. The second task requires the participant to use the calculator to calculate the driving time based on the given distance and maximum speed.

The parameter pl_{mn} recorded the number of attempts that the participant performed to place all markers or tokens correctly. As shown in Table 4, in the first trial, the mean number of attempts is 7.60 in VR and 4.67 in non-VR. This indicates that placing tokens on the map in the VR environment is more challenging than placing markers on a non-VR, 2D interface. The between-version comparison on the first trial for pl_{mn} indicates that there is a significant difference for this parameter, where $F(1,28) = 6.04$, $p = 0.02$. In the second trial, the mean number of attempts in the VR version is reduced more significantly than the number of attempts in the non-VR version, where the mean values are reduced to 4.87 in VR and 4.07 in non-VR. However, there is still a significant difference for pl_{mn}. We observed that when participants were placing tokens in VR, they tended to try repetitively with small moves until reaching the correct position. In non-VR, participants usually moved markers directly to the position that they thought should be correct. Between the two trials, user performance in the second trial was better than user performance in the first trial for both versions of the game. As shown in the "Between-Trial Comparison" columns in Table 4, the ANOVA results are $F(1, 28) = 5.10$, $p = 0.03$ in VR and $F(1, 28) = 4.13$, $p = 0.05$ in non-VR. This means that user performance between the two trials was significantly different.

Table 4. Analysis results of user performance.

Phases	Parameters	Trials	VR Version					Non-VR Version					Between-Version Comparison		
			Mean	Standard Deviation	Between-Trial Comparison			Mean	Standard Deviation	Between-Trial Comparison			$F(1,28)$	p	Significant at $p<0.05$
					$F(1,28)$	p	Significant at $p<0.05$			$F(1,28)$	p	Significant at $p<0.05$			
Planning	pl_{mn}	1	7.60	4.48	5.10	0.03	Yes	4.67	1.11	4.13	0.05	No	6.04	0.02	Yes
		2	4.87	1.35				4.07	0.26				5.04	0.03	Yes
	pl_{mt}	1	03:05	01:24	29.21	<0.01	Yes	01:33	00:36	19.70	<0.01	Yes	15.42	<0.01	Yes
		2	01:04	00:23				00:51	00:06				4.41	0.04	Yes
	pl_{mt1}	1	01:39	00:47	31.20	<0.01	Yes	00:40	00:09	15.22	<0.01	Yes	6.04	0.02	Yes
		2	00:29	00:10				00:31	00:02				0.46	0.50	No
	pl_{ct}	1	01:34	00:50	23.91	<0.01	Yes	01:06	00:27	43.90	<0.01	Yes	3.61	0.07	No
		2	00:30	00:08				00:16	00:10				18.42	<0.01	Yes
Preparing	pr_n	1	6.20	1.61	0.50	0.49	No	6.40	1.88	0.00	1.00	No	0.10	0.70	No
		2	5.87	0.83				6.40	3.44				0.34	0.56	No
	pr_t	1	04:11	01:15	55.60	<0.01	Yes	03:33	01:22	41.68	<0.01	Yes	1.77	0.19	No
		2	01:40	00:23				01:05	00:31				12.03	<0.01	Yes
	pr_{t1}	1	00:40	00:13	56.61	<0.01	Yes	00:37	00:17	27.45	<0.01	Yes	0.25	0.62	No
		2	00:13	00:03				00:13	00:03				0.06	0.81	No
Driving	dr_t	1	05:36	01:24	10.45	<0.01	Yes	05:05	01:01	14.55	<0.01	Yes	1.35	0.26	No
		2	04:09	01:00				04:00	00:24				0.27	0.61	No
	dr_n	1	5.53	2.88	4.52	<0.01	Yes	2.93	3.01	0.00	1.0	No	5.71	0.02	Yes
		2	3.27	2.96				3.00	2.56				0.07	0.79	No

The parameter pl_{mt} recorded the total time the participant spent on placing markers or tokens. In the first trial, the mean was 03:05 in VR and 01:33 in non-VR. Thus, the time in VR was 1.98 times longer than the time in non-VR. In the second trial, the mean was 01:04 in the VR version and 00:51 in non-VR. The time in the VR version was 1.25 times longer than the time in the non-VR version. In each of the trials, the time duration was a significant difference between the two versions. Participants who played the VR version of the game spent a longer time in the planning phase than the participants playing the non-VR version. As shown in the "Between-Trial Comparison" columns in Table 4, the value of pl_{mt} is decreased significantly from the first trial to the second trial in both versions, and such a decrease is significant since $F(1, 28) = 29.21$, $p < 0.01$ in the VR version and $F(1, 28) = 19.70$, $p < 0.01$ in the non-VR version.

The parameter pl_{mt1} recorded the time that the participant spent on placing the first marker or token. The value of this parameter indicates how long the participant took to learn and understand task requirements and basic operations. In the first trial, the mean value of pl_{mt1} in the VR version was 01:39 and the mean value was 00:40 in the non-VR version. The difference in pl_{mt1} between the two versions is significant since the results show $F(1, 28) = 6.04$, $p = 0.02$ in the "Between-Version Comparison" column. This indicates that participants who played the VR version spent longer learning how to operate the game than the participants who played the non-VR version. In the second trial, there was no significant difference in pl_{mt1} between the two versions. The time for the VR version was 00:29 and the time for the non-VR version was 00:31. The value of pl_{mt1} decreased significantly from the first trial to the second trial in both versions. We observed that after the first trial, participants understood clearly the task requirements and basic operations. There was no learning difficulty for participants in the second trial.

The parameter pl_{ct} recorded the time spent on calculating the driving time with the given distance and maximum speed. The results do not show a significant difference for this parameter between the two versions or between the trials. However, we observed that from the first trial to the second trial, the value of this parameter decreased significantly in both versions.

6.1.2. Evaluation of the Preparing Phase

In the preparing phase, participants assembled the lunar rover with constantly proposed requirements for each of the five tasks, as described in Section 4.2. After the participants added all the devices that they thought were required for a task, they clicked the submission button. If the devices were added correctly, they would move to the next task. Otherwise, they had to select appropriate devices again and resubmit. Note that in the game, the order of the five tasks is different in the first trial and the second trial. The order the devices are displayed in is random and so will most likely be different between the two trials. The parameter pr_n records the number of submissions the participant made. The results do not show any significant difference in pr_n between the two versions or between the two trials. However, from the first trial to the second trial, the number of submissions decreased significantly in both versions. This is because the participants became familiar with the devices in the inventory in the second trial, so they could determine the correct answer more quickly.

For the first trial, the total time (pr_t) that the participants spent on finishing this phase in the VR version ($M = 04:11, SD = 01:15$) was not significantly different to the time spent in the non-VR version ($M = 03:33, SD = 01:22$). However, there was a significant difference in the second trial. In the second trial, the mean value of pr_t was 01:40 in the VR version, and the mean value was 01:05 in the non-VR version. Between the two versions, the results have $F(1,28) = 12.03, p < 0.01$. In each version, the value of pr_t decreased significantly from the first trial to the second trial.

The parameter pr_{t1} records the time that the participant spent on the first submission. In each trial, there is not any significant difference between the two versions. In each version, the value of pr_{t1} decreases significantly from the first trial to the second trial.

6.1.3. Evaluation of the Driving Phase

Participants drove through all station stops and came back to the Lunar Module (the landing location). In each trial, the total time that the participant spent on finishing this phase (dr_t) was not significantly different between the two versions, as shown in the "'Between-Version Comparison" column in Table 4. For example, for the first trial, the mean value of dr_t in the VR version was 05:36, which is almost the same as the mean value in the non-VR version, which was 05:05. In each version, the decrease in the value from the first trial to the second trial was significant, as shown in the "Between-Trial Comparison" columns in Table 4 for each of the versions.

In the driving phase, participants had to monitor the battery temperature and motor temperature in order to avoid the issue of overheating. If the rover starts overheating, the participant should reduce the speed or completely stop the rover to let it cool down. We defined the parameter dr_n to record the number of times that the rover overheated. In the first trial, the mean value of dr_n in the VR version was 5.53, which is higher than the value of 2.93 in the non-VR version. The difference in dr_n between the two versions in the first trial was significant ($F(1,28) = 5.71, p = 0.02$). This indicates that in the first trial, the participants tended to drive more aggressively in VR than in the non-VR version of the game. In the second trial, however, the participants who played the VR version of the game performed more calmly than in the first trial. There was no significant difference in dr_n between the two versions in the second trial ($F(1,28) = 0.07, p = 0.79$). We observed that in the second trial of the VR version, participants paid more attentions to the prompted messages than in their first trial.

6.2. Game Engagement

We employed the GEQ and a post-game interview questionnaire to evaluate game engagement. Table 1 shows the results of the GEQ. The GEQ provides a total of 19 questions. The VR version of the game had better results than the non-VR version in 11 questions. For example, for the question "I feel spaced out", the VR version received a score 47% higher than the non-VR version of the game. For the question "I lose track of where I am", the VR version received a score 52% higher than the non-VR version of the game. The VR version of the game also received higher scores on the questions "I feel scared" and "I get wound up". Since they are negative questions, higher scores indicate a weaker engagement in the game.

The results of the interview questionnaire are showed in Table 2. The averaged rating for the non-VR version was 8.26/10, and the averaged rating for the VR version is 8.60/10. The overall rating for the VR version was higher than the non-VR version, but this was not a significant difference. Participants rated the difficulty level of the VR version (4.07/10) higher than the non-VR version (3.13/10).

The interview questionnaire asked participants whether they wanted to learn more about the historical events of lunar exploration missions. The VR version of the game received a rating of 8.47/10, which is significantly higher than the non-VR version's rating of 7.67/10 ($F(1,28) = 4.97$, $p = 0.03$). Moreover, for the question asking whether they gained interest in space science, the VR version received the rating of 8.33/10, which is significantly higher than the rating of 7.60/10 for the non-VR version ($F(1,28) = 4.52$, $p = 0.04$).

Motion sickness could be an issue in the VR version of the game. The interview questionnaire asked participants if they felt motion-sick during play. We received the rating of 2.60/10 for the non-VR version and 3.90/10 for the VR version. Thus, the VR version has a higher chance of making participants feel motion-sick. We observed that some participants started feeling motion-sick when they entered the driving phase because of the bumpy lunar terrain surface. This has a negative impact on VR-based game engagement.

In the game, there is a crater near the landing site that is displayable on the terrain map in the planning phase and approachable in the driving phase. The interview questionnaire asked participants whether or not they noticed the crater and asked them to describe the location of the crater. A total of 16 participants answered correctly, including 6 participants who played the non-VR version and 10 participants who played the VR version. We observed that participants who played the VR version gained a better understanding of the nearby environment and performed better on tasks of identifying locations and directions.

As we observed in the literature, astronauts in the real lunar exploration mission gained information on the mission either during the pre-launch training or through voice communications with HQ. In the game, participants had to read texts on the screen to know the functionality of the scientific instruments or follow along the mission tasks. Reading screen displays is not the intuitive method for this gameplay, and consequently it may decrease the level of engagement.

6.3. Discussion

In our experiment, when participants were playing the VR version of the game, they used extra time to get familiar with the VR environment and learn how to operate the handheld controllers. Subsequently, it took longer for them to finish the game. In contrast, the participants did not need a learning or tutorial session to learn how to use the keyboard and mouse in the non-VR version. In our experiment, we noticed that participants took about 22 min to finish the VR version of the game and 17.5 min to finish the non-VR version. We also found that performing an operation in VR usually takes longer than performing one in the non-VR version. In the non-VR version, the participants moved the mouse and clicked a button to finish an in-game event. However, in the VR version, participants had to point to an object, grab it, and then drag it to a certain place. This may make the participant spend longer on completing a game in VR. In the second trial in the study, participants usually performed better than in the first trial. This happened in both the VR and non-VR versions. User performance improvement is more significant in the VR version, and the performance gap between the two versions was largely reduced in the second trial.

We noticed that even though some participants experienced motion sickness, they still provided positive feedback on their overall gameplay experience. Participants who played the VR version also gave higher scores than the participants who played the non-VR version for the interview question about how well the game promotes interest in space science.

7. Conclusions and Future Work

In this paper, we designed, implemented, and studied a serious lunar exploration mission VR game. We demonstrated the design differences between the VR version and the non-VR version. We discussed the usability and engagement study between VR and non-VR versions. The results of our experiment indicate that in our game, the non-VR version leads to better user performance than the VR version. Users usually needed extra time to get familiar with the VR environment and the use of handheld controllers. After practicing in VR however, the performance gap with the non-VR version reduced significantly. More importantly, the VR version improves the game engagement and enhances the interest of players in learning more about space science and the historical events of lunar exploration.

In the future, we plan to add a multiplayer mode to the game. We want to study the usability and game engagement through multiplayer cooperation on lunar exploration missions. We also plan to improve the naturalness of the interaction with the virtual reality environment. For example, we could employ a hand-gesture-based input modality for gaming controls instead of using handheld controllers.

Author Contributions: Conceptualization, L.C., C.P., and J.T.H.; methodology, L.C., C.P., and J.T.H.; software, L.C. and C.P.; validation, L.C., C.P., and J.T.H.; formal analysis, L.C.; investigation, L.C. and C.P.; resources, C.P. and J.T.H.; data curation, L.C.; writing—original draft preparation, L.C. and C.P.; writing—review and editing, L.C., C.P., and J.T.H.; visualization, L.C.

Funding: This research received no external funding.

Acknowledgments: We gratefully acknowledge the support of the NVIDIA Corporation with the donation of a GPU card used in this project. We thank anonymous reviewers for their comments. The study was performed at the University of Alabama in Huntsville (UAH), USA. The Institutional Review Board (IRB) application (E201922) for the study was submitted and approved by the UAH Institutional Review Board of Human Subjects Committee. We thank the participants for their participation in the study. We also thank the Center for Media, Arts, Games, Interaction & Creativity (MAGIC) at the Rochester Institute of Technology. The lunar rover model was obtained from Hameed (https://www.deviantart.com/hameed/art/Lunar-Rover-Downloadable-3D-Model-With-Textures-439471995) for non-commercial use.

Conflicts of Interest: The authors declare no conflict of interest.

References

1. Schuemie, M.J.; Van Der Straaten, P.; Krijn, M.; Van Der Mast, C.A. Research on presence in virtual reality: A survey. *CyberPsychol. Behav.* **2001**, *4*, 183–201. [CrossRef] [PubMed]
2. Susi, T.; Johannesson, M.; Backlund, P. *Serious Games: An Overview*; IKI Technical Reports; School of Humanities and Informatics, University of Skövde: Skövde, Sweden, 2007. Available online: http://urn.kb.se/resolve?urn=urn:nbn:se:his:diva-1279 (accessed on 29 June 2019).
3. Djaouti, D.; Alvarez, J.; Jessel, J.P. Classifying serious games: The G/P/S model. In *Handbook of Research on Improving Learning and Motivation through Educational Games: Multidisciplinary Approaches*; IGI Global: Hershey, PA, USA, 2011; pp. 118–136.
4. Peng, C.; Cao, L.; Timalsena, S. Gamification of Apollo lunar exploration missions for learning engagement. *Entertain. Comput.* **2017**, *19*, 53–64. [CrossRef]
5. Brockmyer, J.H.; Fox, C.M.; Curtiss, K.A.; McBroom, E.; Burkhart, K.M.; Pidruzny, J.N. The development of the Game Engagement Questionnaire: A measure of engagement in video game-playing. *J. Exp. Soc. Psychol.* **2009**, *45*, 624–634. [CrossRef]
6. Mikropoulos, T.A.; Natsis, A. Educational virtual environments: A ten-year review of empirical research (1999–2009). *Comput. Educ.* **2011**, *56*, 769–780. [CrossRef]
7. Potkonjak, V.; Gardner, M.; Callaghan, V.; Mattila, P.; Guetl, C.; Petrović, V.M.; Jovanović, K. Virtual laboratories for education in science, technology, and engineering: A review. *Comput. Educ.* **2016**, *95*, 309–327. [CrossRef]

8. Alhalabi, W. Virtual reality systems enhance students' achievements in engineering education. *Behav. Inf. Technol.* **2016**, *35*, 919–925. [CrossRef]
9. Jensen, L.; Konradsen, F. A review of the use of virtual reality head-mounted displays in education and training. *Educ. Inf. Technol.* **2018**, *23*, 1515–1529. [CrossRef]
10. Freina, L.; Ott, M. A literature review on immersive virtual reality in education: State of the art and perspectives. In Proceedings of the International Scientific Conference eLearning and Software for Education, Bucharest, Romania, 23–24 April 2015; Volume 1, p. 133.
11. Virvou, M.; Katsionis, G. On the usability and likeability of virtual reality games for education: The case of VR-ENGAGE. *Comput. Educ.* **2008**, *50*, 154–178. [CrossRef]
12. Mei, H.H.; Sheng, L.S. Applying situated learning in a virtual reality system to enhance learning motivation. *Int. J. Inf. Educ. Technol.* **2011**, *1*, 298–302.
13. Cheng, A.; Yang, L.; Andersen, E. Teaching language and culture with a virtual reality game. In Proceedings of the 2017 CHI Conference on Human Factors in Computing Systems, Denver, CO, USA, 6–11 May 2017; pp. 541–549.
14. Adamo-Villani, N.; Wilbur, R.B. Effects of platform (immersive versus non-immersive) on usability and enjoyment of a virtual learning environment for deaf and hearing children. In Proceedings of the Eurographics Symposium on Virtual Environments (EGVE 2008), Eindhoven, The Netherlands, 29–30 May 2008; pp. 8–19.
15. Parmar, D.; Bertrand, J.; Babu, S.V.; Madathil, K.; Zelaya, M.; Wang, T.; Wagner, J.; Gramopadhye, A.K.; Frady, K. A comparative evaluation of viewing metaphors on psychophysical skills education in an interactive virtual environment. *Virtual Real.* **2016**, *20*, 141–157. [CrossRef]
16. Greenwald, S.W.; Corning, W.; Funk, M.; Maes, P. Comparing Learning in Virtual Reality with Learning on a 2D Screen Using Electrostatics Activities. *J. UCS* **2018**, *24*, 220–245.
17. Olmos-Raya, E.; Ferreira-Cavalcanti, J.; Contero, M.; Castellanos-Baena, M.; Chicci-Giglioli, I.; Alcañiz, M. Mobile virtual reality as an educational platform: A pilot study on the impact of immersion and positive emotion induction in the learning process. *Eurasia J. Math. Sci. Technol. Educ.* **2018**, *14*, 2045–2057. [CrossRef]
18. Zizza, C.; Starr, A.; Hudson, D.; Nuguri, S.S.; Calyam, P.; He, Z. Towards a social virtual reality learning environment in high fidelity. In Proceedings of the 2018 15th IEEE Annual Consumer Communications & Networking Conference (CCNC), Las Vegas, NV, USA, 12–15 January 2018; pp. 1–4.
19. Buttussi, F.; Chittaro, L. Effects of different types of virtual reality display on presence and learning in a safety training scenario. *IEEE Trans. Vis. Comput. Graph.* **2017**, *24*, 1063–1076. [CrossRef] [PubMed]
20. Bowman, D.A.; Gabbard, J.L.; Hix, D. A survey of usability evaluation in virtual environments: Classification and comparison of methods. *Presence Teleoper. Virtual Environ.* **2002**, *11*, 404–424. [CrossRef]
21. Sutcliffe, A.G.; Kaur, K.D. Evaluating the usability of virtual reality user interfaces. *Behav. Inf. Technol.* **2000**, *19*, 415–426. [CrossRef]
22. McMahan, R.P.; Bowman, D.A.; Zielinski, D.J.; Brady, R.B. Evaluating display fidelity and interaction fidelity in a virtual reality game. *IEEE Trans. Vis. Comput. Graph.* **2012**, *18*, 626–633. [CrossRef] [PubMed]
23. Sharples, S.; Cobb, S.; Moody, A.; Wilson, J.R. Virtual reality induced symptoms and effects (VRISE): Comparison of head mounted display (HMD), desktop and projection display systems. *Displays* **2008**, *29*, 58–69. [CrossRef]
24. Abdul Jabbar, A.I.; Felicia, P. Gameplay engagement and learning in game-based learning: A systematic review. *Rev. Educ. Res.* **2015**, *85*, 740–779. [CrossRef]
25. Oksanen, K. Subjective experience and sociability in a collaborative serious game. *Simul. Gaming* **2013**, *44*, 767–793. [CrossRef]
26. Byun, J.; Loh, C.S. Audial engagement: Effects of game sound on learner engagement in digital game-based learning environments. *Comput. Hum. Behav.* **2015**, *46*, 129–138. [CrossRef]
27. Liu, M.; Lee, J.; Kang, J.; Liu, S. What we can learn from the data: A multiple-case study examining behavior patterns by students with different characteristics in using a serious game. *Technol. Knowl. Learn.* **2016**, *21*, 33–57. [CrossRef]
28. Hookham, G.; Nesbitt, K.; Kay-Lambkin, F. Comparing usability and engagement between a serious game and a traditional online program. In Proceedings of the Australasian Computer Science Week Multiconference, Canberra, Australia, 1–5 February 2016; p. 54.

29. Fernandez-Cervantes, V.; Stroutia, E. Virtual-Gym vR: A Virtual Reality Platform for Personalized Exergames. In Proceedings of the 2019 IEEE Conference on Virtual Reality and 3D User Interfaces (VR), Osaka, Japan, 23–27 March 2019; pp. 920–921.
30. McMahan, R.P. Exploring the Effects of Higher-Fidelity Display and Interaction for Virtual Reality Games. Ph.D. Thesis, Virginia Tech, Blacksburg, VA, USA, 2011.
31. National Science Foundation. Percentage Distribution of Bachelor's Degrees Awarded, by Major Field Group: 1966–2010. 2013. Available online: https://www.nsf.gov/statistics/nsf13327/pdf/tab6.pdf (accessed on 14 August 2019).
32. Markovich, S.J.; Chatzky, A. Space Exploration and U.S. Competitiveness. 2019. Available online: https://www.cfr.org/backgrounder/space-exploration-and-us-competitiveness (accessed on 14 August 2019).
33. Johnson, R.T.; Johnson, D.W. Active learning: Cooperation in the classroom. *Annu. Rep. Educ. Psychol. Jpn.* **2008**, *47*, 29–30. [CrossRef]
34. NASA's Jet Propulsion Laboratory. NASA's Eyes. 2010. Available online: https://eyes.nasa.gov/ (accessed on 14 August 2019).
35. NASA's Jet Propulsion Laboratory. Space Place: Explore Earth and Space. 2019. Available online: https://spaceplace.nasa.gov/ (accessed on 16 August 2019).
36. NASA. NASA 3D Resources. 2019. Available online: https://nasa3d.arc.nasa.gov/ (accessed on 16 August 2019).
37. Antoun, C. Station Spacewalk Game. 2010. Available online: https://www.nasa.gov/multimedia/3d_resources/station_spacewalk_game.html (accessed on 16 August 2019).
38. NASA's Jet Propulsion Laboratory and Goddard Space Flight Center. Earth's Moon. 2019. Available online: https://moon.nasa.gov/ (accessed on 16 August 2019).
39. Virtual Reality Education & Corporate Training. Apollo 11 VR. 2018. Available online: https://immersivevreducation.com/apollo-11-vr/ (accessed on 16 August 2019).
40. Naval Aviation Museum. Apollo 11 Virtual Reality. 2018. Available online: https://www.navalaviationmuseum.org/attractions/apollo-11-virtual-reality/ (accessed on 16 August 2019).
41. Cao, L.; Peng, C.; Hansberger, J. A Large Curved Display System in Virtual Reality for Immersive Data Interaction. In Proceedings of the 2019 IEEE Games, Entertainment, Media Conference (GEM), New Haven, CT, USA, 18–21 June 2019.

© 2019 by the authors. Licensee MDPI, Basel, Switzerland. This article is an open access article distributed under the terms and conditions of the Creative Commons Attribution (CC BY) license (http://creativecommons.org/licenses/by/4.0/).

Article
A Guide for Game-Design-Based Gamification

Francisco J. Gallego-Durán *, Carlos J. Villagrá-Arnedo, Rosana Satorre-Cuerda,
Patricia Compañ-Rosique, Rafael Molina-Carmona and Faraón Llorens-Largo

Cátedra Santander-UA de Transformación Digital, Universidad de Alicante, 03690 San Vicente del Raspeig, Spain; villagra@ua.es (C.J.V.-A.); rosana.satorre@ua.es (R.S.-C.); patricia.company@ua.es (P.C.-R.); rmolina@ua.es (R.M.-C.); Faraon.Llorens@ua.es (F.L.-L.)
* Correspondence: fjgallego@ua.es

Received: 8 July 2019; Accepted: 24 October 2019; Published: 5 November 2019

Abstract: Many researchers consider Gamification as a powerful way to improve education. Many studies show improvements with respect to traditional methodologies. Several educational strategies have also been combined with Gamification with interesting results. Interest is growing and evidence suggest Gamification has a promising future. However, there is a barrier preventing many researchers from properly understanding Gamification principles. Gamification focuses of engaging trainees in learning with same intensity that games engage players on playing. But only some very well designed games achieve this level of engagement. Designing truly entertaining games is a difficult task with a great artistic component. Although some studies have tried to clarify how Game Design produces fun, there is no scientific consensus. Well established knowledge on Game Design resides in sets of rules of thumb and good practices, based on empirical experience. Game industry professionals acquire this experience through practice. Most educators and researchers often overlook the need for such experience to successfully design Gamification. And so, many research papers focus on single game-elements like points, present non-gaming activities like questionnaires, design non-engaging activities or fail to comprehend the underlying principles on why their designs do not yield expected results. This work presents a rubric for educators and researchers to start working in Gamification without previous experience in Game Design. This rubric decomposes the continuous space of Game Design into a set of ten discrete characteristics. It is aimed at diminishing the entry barrier and helping to acquire initial experience with Game Design fundamentals. The main proposed uses are twofold: to analyse existing games or gamified activities gaining a better understanding of their strengths and weaknesses and to help in the design or improvement of activities. Focus is on Game Design characteristics rather than game elements, similarly to professional game designers. The goal is to help gaining experience towards designing successful Gamification environments. Presented rubric is based on our previous design experience, compared and contrasted with literature, and empirically tested with some example games and gamified activities.

Keywords: gamification; game design; rubric

1. Introduction

In recent years, Gamification [1,2] is getting considered a magic solution for most educational problems. Many researchers and practitioners chase it, and many studies try to unveil its secrets and details. In one form or another, the term and the field are acknowledging the power of games to engage and induce states of flow in players. Gamification chases this power to apply it to environments that originally are not ludic. The aim is to get people engaged in serious or important work with the same intrinsic motivation than in games.

This enterprise is noble but extremely complicated. As more and more research is being carried out, results remain unclear [3–8]. Hundreds of research experiences have been undertaken with mixed results. Many studies find benefits when applying Gamification, but many others do not and even some of them report damage. Overall tendency seems to report some small but measurable benefits. These results are quite unexpected compared to the exponential rise in game sales and gaming culture in general.

The problem with most Gamification research seems to be in its different focus from actual Game Design. Many studies pursue scientific isolation of statistical variables. This leads them to consider the isolated influence of individual game elements like points, badges and leaderboards in motivation and behaviour change. The problem with this approach is that a game is not an unrelated set of game elements. Metaphorically, a game is similar to a grand-cuisine dish: testing its isolated ingredients in other contexts does not convey useful information to learn to cook the dish.

This view is supported by relevant Gamification practitioners like Kevin Werbach, Yu-kai Chou or Sebastian Deterding [9–11] and also Game Design experts like Raph Koster or Jesse Schell [12,13]. In Werbach's words [9]: "Clearly not everything that includes a game element constitutes gamification. Examinations in schools, for example, give out points and are non-game contexts." Deterding goes beyond that in Reference [11]: "The main task of rethinking Gamification is to rescue it from the gamifiers." For Deterding, the majority of gamifiers are confused as they simple try to add points, badges and leaderboards to everything, with great disregard to the complexities of Game Design.

Games are complex environments that deliver experiences to players [13]. They are made of game elements, similar to a dish is made of ingredients but the process, interactions, uses and objectives are key for the final result:

> "Gamification should be understood as a process. Specifically, it is **the process of making activities more game-like**. Conceiving of Gamification as a process creates a better fit between academic and practitioner perspectives. Even more important, it focuses attention on the creation of game-like experiences, pushing against shallow approaches that can easily become manipulative. A final benefit of this approach is that it connects Gamification to persuasive design." Kevin Werbach [9]

These reasons could explain why there is no scientific consensus on a formal approach to Gamification. There are analyses of the characteristics of good games [14,15] which Gamification pursues. There also are methodological approaches, design frameworks and even descriptions of design patterns based on Game Design principles, good practices and experience [3,16–21]. However, all approaches rely on subjective interpretation and creative design. In fact, many professional Game Designers and researchers express their view that games cannot be formally specified at all [22,23].

Even if games cannot be formally specified and individual game element research does not yield complete information, there are useful approaches [16,24]. Assuming that Game and Gamification Design are artistic in essence, approaches the focus on acquiring design experience. There is no need to solve "what-is-a-game" philosophical debate. Game-like designs able to become engaging voluntary experiences for players could be successful. Willingness can make experiences fall in persuasive or seductive sides of Tromp, Hekkert and Verbeek's matrix in which design can influence behaviour [25]. Similar to References [9,11–13], this work focuses on this practical approach.

The main goal of this work is to help acquire Game Design experience for Gamification. Experienced practitioners may found methods, frameworks or models cited in the literature more suitable to their needs, particularly References [3,17,18,20,26]. These works have great value but require previous Game-Design-Based Gamification experience to be fully comprehended and put into practice. To build such required previous experience a practical and simple approach is proposed: a measurement tool, a rubric, with great focus on Game Design aspects rather than on game elements.

Acquiring Design Experience

As previous design experience seems key [9,12,13], our proposal for new practitioners is creating and testing their own designs. In our experience, iterating over own designs leads to obtaining solid game-design-based Gamification design skills. However, analysing and improving designs results in an almost impossible task for inexperienced designers. On the absence of personal experience to rely on, the only valid source is testing. Testing with trainees is essential but doing so with no previous design guidance could result on a extremely slow and frustrating discovery process. This is an entry barrier that can produce two important problems: too many failures on initial attempts, and abandon due to frustration. Moreover, when initial failures are not identified as a consequence of lack of experience, they can result in research papers blaming the field itself.

During fifteen years teaching Game Development and Gamification [27,28], we have perceived a great difficulty to pass on design experience to new practitioners. The problem, as discussed, seems to be on the artistic nature of Game Design. Novice practitioners often underestimate the complexity of creating a design that can be put in practice, not to say a successful one. This is problematic, as their initial experiences will probably fail and be frustrating. There are design frameworks, methods and guidelines proposed for game and Gamification [3,16,18–21,29] that could help in creating better first designs. However, these proposals are either general or specifically for experts. They are not designed with novices in mind and can easily result overwhelming for them. For instance, Kreimeier's patterns [16] condense many designers' experiences. This is highly valuable but almost impossible to properly understand without previous experience on pitfalls and failures. Tondello et al. [20] explicitly state "Our set of heuristics is aimed at enabling experts to identify gaps in a gameful system's design" which clearly leaves novices out. Linehan et al. [3] propose to use Applied Behaviour Analysis from the field of psychology with many interesting theoretical explanations. This is too much theoretical information for novices which probably will require several testing iterations to relate it to actual practice. Similarly, Self-Determination Theory (SDT) [29] is the most widely cited theoretical framework. In essence, SDT is easy to understand but too generic. Novices need more specific and game related descriptions, as SDT is purely psychological. Hunicke et al. proposal [18] splits Game Design into three blocks: Mechanics, Dynamics and Aesthetics (MDA). This simple classification helps organizing designs, which is very useful for novices but does not help in measuring their value, comparing with others or giving hints on how to improve them.

To help in this process, this work proposes a game-design-based rubric. This rubric focuses on measuring how well designed a game or activity is, from a game-design-based Gamification perspective. The measure is formalized as a score from 0 to 20 points; the greater the score, the better. The rubric is based on a set of ten characteristics related to successful designs. These ten characteristics have been selected from our previous experience in Game Design and Gamification, partially in accordance to previously discussed works and with an aim to simplify analysis. The goal is being useful enough to serve as analysis and design tool, at the same time as being simple enough to help novices.

It is important to remember that there is no known way to perform an objective assessment of a given design. The assessment obtained with the proposed rubric is to be considered a simplified initial guidance. This guidance is targeted at inexperience designers to help them overcome the entry barrier. In this sense, the rubric helps discretizing designs and moving them from the artistic to the analytic dimension. It also helps identifying potential areas for improvement, pointing to those underperforming characteristics of a given design. These values are complementary to previously analyzed formal tools and frameworks, which makes it interesting to be combined with them.

Section 2 describes the ten selected characteristics for the rubric in detail, explaining their design implications. Section 3 presents the rubric and explains its design constraints and criteria. Section 4 shows some initial evidence on the validity of the rubric by applying it to four activity samples: a commercial game, a gamified course from literature, a learning activity and a gamified version of the learning activity. Finally, Section 5 sums up conclusions and limitations of this work.

2. Ten Relevant Characteristics for Game-Design-Based Gamification

This section describes in detail the ten selected characteristics in the proposed rubric. The importance of each one is discussed by highlighting relevant psychological arguments considered and comparing classical educative environments with successful commercial games. Moreover, considerations from prestigious game designers are also cited and analysed from their published works.

This set of characteristics greatly overlaps those described by previous works [3,16,17,19–21], specially Gee's learning principles good games incorporate [14,15]. Most works are generalist and some are specifically focused on experts. That could be overwhelming for novices. Our proposed set aims to fill in this gap. Its intended value comes from its use case goal: simple and easy to understand for new practitioners.

We acknowledge that this set is subjective in nature, despite the arguments presented and discussed. We propose them out of previous Game Design and Gamification experience and we expect following works to help refining it after gathering appropriate usage evidence. Section 4 gives an initial piece of evidence on the validity of this set for guidance and ground basis to start retrieving more evidence.

Description of the ten characteristics follows in no particular order.

2.1. Open Decision Space

Autonomy is one of the center points of intrinsic motivation. In order for a trainee to be truly autonomous, there must exist different possible decisions to take. In fact, the greater the space of possible decisions over time, the better. However, there are some common misconceptions whose analysis is relevant.

- Correct decisions ill-form decision spaces.
  Many Gamification designs reside on questions or alternative paths, being only one of them correct. This represents and ill-formed decision space because there is no true decision to take. Trainees are not being asked to decide and progress but are being tested instead. For an environment to foster autonomy and provide a truly open decision space, decisions should not be designed as correct/incorrect. In contrast, decisions must produce consequences and trainees should be free to play with situations, environments and consequences, experimenting and learning from results.
- Discrete custom-designed decision spaces challenge autonomy.
  It is common to manually design all possible decision choices. It seems natural to attempt to directly transmit knowledge to trainees. We teachers tend to transform our knowledge into possible situations, producing some form of decision-tree. Decision space is reduced to pre-designed knowledge, what challenges autonomy. Trainees usually imagine decisions they would take but have to accommodate to designed choices. Creativity is prevented, curiosity diminishes and frustration raises. Open decision spaces that let trainees experiment with their ideas tend to be continuous, not pre-designed, more similar to simulations than to decision trees.
- Failing to consider movements and interactions as decisions.
  The term 'decision' is naturally related to high-level abstract thinking and neo-cortex processing. But any action a trainee performs on any instant is a decision. Many designs do not consider them as part of the decision space. This produces designs where either trainees cannot move or their movements are meaningless. As interacting with the world is one of our richest sources of information, failing to consider it greatly limits decision spaces.

These points are clearly addressed by great games. Decision spaces are usually continuous, as players can move freely over time and experiment the consequences of their interaction decisions. Take for instance *Super Mario Bros* for *Nintendo Entertainment System* (NES) (see Figure 1) [30]. When facing the first enemy there are virtually infinite ways to do it. We often simplify it by thinking that you can jump on it, jump over it or collide with it and die. However, there are virtually infinite alternatives: jumping earlier or later, higher or lower, faster or slower and so forth. Players could even jump several times back and forth, advance and retreat, or do anything they can imagine based on the free will given by the rules of the world. In fact, Demain et al. prove that *Super Mario Bros* is

PSPACE-complete [30], the harder class of problems that can be solved in polynomial time. Of course, this great complexity comes exactly from the openness of its decision space. Generally speaking, broader decision spaces that are not ill-formed produce more complex problems. And the greater the complexity, the more the options for creative behaviour, which fosters player autonomy.

Figure 1. Start of the first level of *Super Mario Bros* game. There are virtually infinite possible decisions, in the form of movement sequences. Players can press up to 60 combinations of inputs per second.

2.2. Challenge

Designing challenging activities is a key point in Gamification and a difficult task to accomplish. An activity is considered challenging when it tests the limits of our ability in subtle ways. Oversimplifying, a design space can be considered with only two dimensions: difficulty of the task and ability of the trainee. When both difficulty and ability match, trainees are faced with activities that they are able to solve [31]. However, when difficulty is much higher than ability, trainees usually get frustrated. On the contrary, if ability is much higher than difficulty, then trainees probably become bored. There is a narrow space in between both extremes where difficulty and abilities are evenly matched. This simple analysis on activity design-space for challenge is the basis for the theory of the channel of flow [32] (see Figure 2).

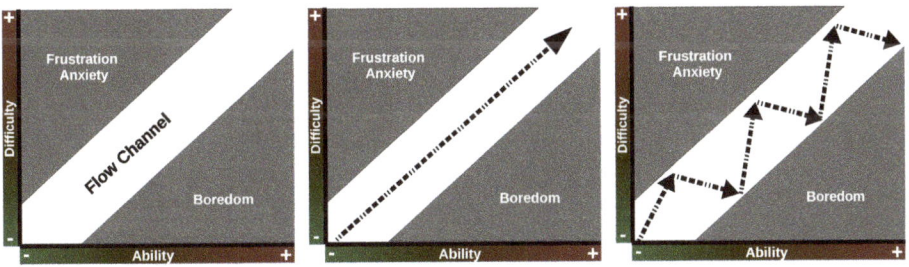

Figure 2. (**Left**) The flow channel. (**center**) Linear incremental difficulty design that perfectly matches abilities. (**right**) Rhythmic incremental difficulty design.

In essence, challenging trainees consists in assigning then interesting tasks that lie on the verge of their abilities. Although simple in concept, challenging trainees in an educational environment is difficult. A trainee can fail in a challenging task several times while learning. Educational environments tend to punish failure by diminishing trainees marks. This is contrary to using challenge as a driver for learning and motivation. In the presence of punishments, trainees avoid difficulty even at the cost of boredom and diminished learning. Marks are the most important outcome pursued by trainees and that must be taken into account.

The flow channel shown in Figure 2 (left) represents a dynamic space. Trainees' abilities evolve over time. As abilities increase, previous challenges become boring and new ones are required. Tasks have to be designed with incremental difficulty in mind. Figure 2 (center) represents the general concept of incremental difficult with an ideal linear progression. However, this progression should not be considered ideal. As we humans are not machines, our brains usually distaste flat linear progressions. Hollywood movie makers usually design movies following a sinusoidal pattern of fast action events followed by relaxed moments. Figure 2 (right) shows this same concept applied to incremental difficulty. This approximation generates stress spikes followed by more relaxed moments. Stress spikes on the verge of the flow channel force trainees to push their limits, adapt and learn. Easier activities let trainees reinforce their sense of progress at the same time they release previous stress and prepare for next spike. Moreover, relaxed moments represent also a psychological reward, as trainees subconsciously acknowledge their new abilities. This pattern is completely similar to General Adaptation Syndrome described by Hans Selye [33] and summarized in Figure 3.

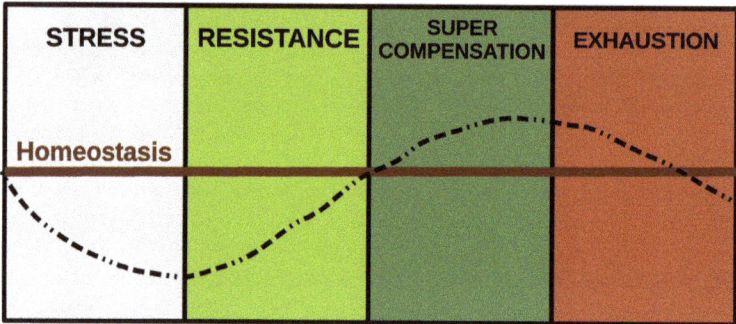

Figure 3. General Adaptation Syndrome. Challenging tasks produce stress, test available abilities and yield failures. Present abilities are improved as resistance, then new ones are developed as super-compensation. Relaxing helps fixing acquired abilities, as continued stress ends up in exhaustion.

2.3. Learning by Trial and Error

This is arguably one of the key differences between great Computer Games and traditional teaching environments. As discussed before, great computer games appropriately challenge players. Proper challenges put players into an edge where they can narrowly succeed or fail depending on their abilities [15,26]. This produces engagement on the most relevant thing a game conveys: learning. Players fail many times and try again, learning from their failures. They continue trying as long as they feel competent to learn and eventually succeed. This guiding force comes from human natural desire to learn and so it reinforces autonomy and will. This cycle happens so naturally because great computer games are safe environments for failure. Players do not want to fail but they are not afraid to do so either. They assume failure as part of the learning process and then try different things to improve their abilities.

In contrast, traditional learning environments are designed to prevent, prosecute and punish failure. This situation often arises from assessing learning on the basis of task results. Two trainees solving a given task can obtain different marks depending on their failures during respective solving processes. This situation drives trainees to focus on preventing failure at all costs to preserve their marks. All learning through challenge and experimentation gets removed from the environment.

In order to mimic computer games and get the benefits from challenge and experimentation, trial and error must be considered as a center way to learn. A proper Gamification design should focus trainees on goals and let them freely experiment and fail without punishment. Trainees need to acknowledge that failing is safe to be confident enough as to experiment. In fact, it would even be better if the environment encourages them to fail and analyse: learning from failure is extremely valuable and often forgotten due to too much focus on task results.

Moreover, solving problems by trial and error, creating solutions, failing, redoing and refining produces "professional experience". In fact, professionals usually say that the greatest expert is a person who has committed all possible mistakes. A great Gamification design understands the importance of experience and designs situations for trainees to learn by trial and error.

2.4. Progress Assessment

Computer games generate virtual environments that evolve with player interactions over time. This evolution immediately informs players about their progress inside the game. Many computer games also include progress measures and feedback systems that constantly inform players about their statistics, achievements, awards, status and, in general sense, progress. Part of the engagement of players in games comes from their sense of progress [3]. Players build upon their own progress as their achievements encourage them to pursue next steps far beyond.

Many learning environments feel very different. Trainees attend lessons and then have to practice or study contents on their own. There is few or none feedback on their progress. Occasionally they can check if they manage to solve some exercises. However, this is radically different from having a constant feedback on progress and clear goals to next levels. One key element in this feedback are measures of already achieved success. Whenever players obtain any award or finish levels, they move on in their gaming adventure and their previous success is acknowledged. Their achievements are never removed, even if they fail afterwards [14]. Compare this with lessons in which a trainee can solve all proposed exercises but fail on the final exam. There is no progress at all because there is no assessment and acknowledgement. All that matters is the result of the final exam. And that is the reason why many trainees do not care about solving exercises during lessons. They only need to prepare them at the end and do well on their exam: progress does not matter at all.

In order to generate engagement and maintain interest, Gamification designs should include one or several forms of progress assessment. Moreover, designing for progress assessment also helps better designing incremental difficulty for challenge, as progress and difficulty are closely related [26]. Ideally, progress assessment should not be based on extrinsic rewards like points or badges. Intrinsic motivation requires trainees to be focused on learning goals per se. Too much emphasis on extrinsic rewards can change trainees' focus, which would be detrimental for learning. For example, some trainees focus on passing subjects to get a degree without worrying about learning. Getting a degree is an extrinsic reward that eclipses their interest for learning and getting abilities. Consequently, using extrinsic rewards to assess progress should be done with care, ensuring that the main focus is always placed on learning goals.

2.5. Feedback

The most relevant difference between computer games and traditional learning environments lays in quantity, quality and rapidity of feedback response. Many good computer games act as simulations, which confers them similar properties to reality: players get immediate feedback response to any of their actions. This is a key point both for learning and engaging: immediate feedback. It is better understood with an example: imagine a child learning to play soccer. Every time the child kicks the ball, it reacts and moves depending on the kick. This feedback lets the child learn applied physics: the child learns to control movement, spin, momentum and force transmitted to the ball through kicking. Now imagine a delayed ball that reacts 24 h after being kicked. Learning how to kick the ball and get a desired reaction would require great patience and effort. Many trainees would rapidly desist, demotivated by such slowness, unable to effective learn. Appropriate, on-time feedback is crucial for both learning and engaging [3,12,13,26].

This is an important problem of many traditional learning environments. Many of the learning activities require to be assessed by a teacher. For instance, trainees solving math exercises wait until they receive teacher corrections. This is similar to the 24-h delayed ball of our previous example. A computer game designed this way would probably be played by no one. A designer would probably envision something more dynamic like "Sum Totaled" mini-game inside "Brain Age Express: Math" from *Nintendo DSi* [34] (see Figure 4). In this mini-game, monsters attack the player who can destroy them by adding the numbers on their bodies. The player has 3 lives that are lost every time a monster hits player's avatar. The activity is based on adding numbers but its rules make it dynamic and the player gets constant, immediate feedback: enemies explode when the player writes down a correct answer and action continues uninterrupted otherwise.

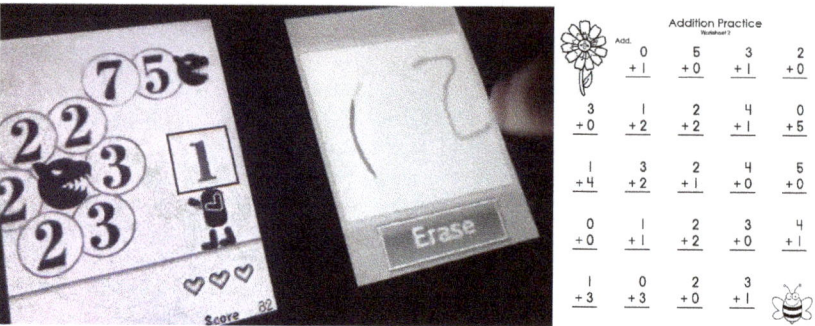

Figure 4. (**Left**) Brain Age 2. Sum Totaled. Player writes '12' (7 + 5) to destroy the monster on the top before it fells down and takes one of the three lives. (**Right**) Traditional addition practice.

Comparison between *Sum Totaled* game and the traditional addition practice in Figure 4 shows the importance of appropriate feedback. Both activities are mathematically the same (apart from their difference in difficulty) but trainees performing it traditionally will have no feedback stimuli to learn from. They will need to wait for teacher corrections. Moreover, game dynamics encourages trainees stop fearing failure and produce more answers, because being quick is crucial. This promotes learning from failure and has the potential to make learning more time/cost efficient.

2.6. Randomness

Randomness is a relevant factor for learning and engaging and links both of them together. In its most fundamental definition, learning is about discovering and modelling patterns and testing constructed models against reality through experimentation. This describes an iterative process for learning, in which engaging arises naturally when trainees constantly find new ways to refine and validate their mental models. An appropriate degree of randomness can keep trainees iterating longer, as their minds will continuously try to refine their models based on unexpected observations. Human minds are not well suited for dealing with probabilities, as they tend to model in terms of strong cause-effect relationships. This generally explains contexts like gambler's fallacy or the hot hand fallacy [35]. These fallacies show us that randomness itself can be used to produce engagement. Therefore, it is quite relevant for games and Gamification designs.

Well designed randomness can provide a useful consequence: surprise. Surprise is one of the most desirable feelings both in learning and playing. Schell describes it as "so basic that we can easily forget about it" in Reference [13]. Schell also describes fun as "pleasure with surprises" and remembers that "Surprise is a crucial part of all entertainment - it is the root of humour, strategy and problem solving. Our brains are hard-wired to enjoy surprises". In fact, surprise happens when observations are radically different from our mental models. The more unexpected the event that happens, the more information it carries: this is a natural consequence of the definition of entropy in information theory by Shannon [36]. This means that surprises are great sources of new information, which can potentially push trainees to revise their mental models, that is, to learn.

Consequently, it is key to consider appropriate uses of randomness in our Gamification designs to foster engagement and learning through surprise.

2.7. Discovery

As Koster states in Reference [12] a good game is one that "keeps the player learning". One of the most important ways to keep players learning is presenting them with new content at an adequate rate. This renews interest in the game and keeps players eager to continue discovering more. It can also trigger surprise, depending on the nature of the new content and the way it is presented. However, discovery is as difficult to design as challenge. New content has to build up on previous content to balance novelty and familiarity. Similar to the flow channel (Figure 2), if a content is radically new it can easily be difficult to understand or accept. New information that cannot link to pre-existing mental models becomes similar to noise: no pattern can be found in the information and so it cannot be modelled and learnt. Some degree of familiarity is needed to help players understand, accept and enjoy new content but too much would eclipse novelty, making new content not feel new at all.

Games present basically two ways for new content delivery: discovery and unlocking. Unlocking works by asking players to perform some achievements to unlock new content. Typically, this means finishing some levels before being able to play new ones. The other way is by placing new content in such a way that players will discover it while playing. Discovery can be equivalent to unlocking by delivering same content at same rate. However, well designed discovery can produce better feelings on players, like surprise, reaffirmation and self-esteem raise. Moreover, discovery can also be designed non-linearly. Games can have secret content, not required to succeed, but present only for players that go beyond normal play. This is also an indirect way to reward players for their attention to detail, research or clever play. It is also an interesting way to convey rewards, as discovery would not be perceived as a reward but as a personal achievement. This has higher probability to foster intrinsic motivation.

Discovery is not commonly used in Gamification. This is probably due to the difficulty of content and activity design. Educational contexts tend to be linear and content is usually known beforehand. Trainees expect contents to be introduced first, then explained linearly. This relates to what we stated in Section 2.1: activities and contents are usually designed with a single correct path, expecting a concrete answer. To add discovery, designs require open spaces in which trainees decisions are relevant. Otherwise, discovery has no meaning at all. And this is the root of the difficulty for including discovery: it is difficult to design activities or content with proper open decision spaces. So, when willing to introduce discovery, it is advisable first to think about activities with open decision spaces.

2.8. Emotional Entailment

Emotions have been generally ignored in education. Educational context usually focus on factual content, leaning and methodologies for better learning. Everything tends to be concentrated on effectiveness of factual learning. Emotions are seldom considered. However, it is quite common for trainees to define their teachers in function of their feelings. Usual comments include "I like lessons from this teacher because they are X", being X a qualifier like "fun", "entertaining", "approachable", "kind". Emotions are a key factor in all human relations and they also play a key role in learning and engaging. It is known that high intensity emotions produce long lasting memories. In fact, even before the term Gamification were coined, many studies targeted fun as a catalyst for learning [37–39].

Similar to movies, games cannot be successful without paying close attention to emotions. At the very least, a game is always expected to be fun. But fun, like any other emotion, is created inside player's mind. In Schell's words, "When people play games, they have an experience. It is this experience that the designer cares about. Without the experience, the game is worthless" [13]. The game itself is not the experience rather than the tool that enables it. That is what makes anything we create different for each person: the experience happens always in the mind of the player. And that makes it so important for games to be emotionally entailing, because they will attach to emotions in the mind of the player producing a much better and personal experience.

Gamification tends to use the same main tools games use to construct emotional entailment: characters, stories and aesthetics [13]. The problem usually lies in the complexity of these tools. All of them require great abilities and long periods of time when trying to mimic what commercial games do. This is too expensive and usually not cost/effective in educational environments. Simple approaches are preferred in this case: simple stories like "escape from enchanted house" or "disarm the bomb" could be enough for an emotional entailing environment. Trainees could be given freedom to create their own characters (like in role playing games) and aesthetics could be imaginary. Moreover, direct interaction between trainees could also help creating emotional entailment. Forming groups, sharing challenges and achieving common goals are preferred approaches in educational environments.

2.9. Playfulness Enabled

A playfulness enabled game refers to its versatility to be used as a toy. Games have goals, toys do not. Some game can be played without focusing on goals, in a playful way. Some examples include Minecraft, Grand Theft Auto series or Goat Simulator (see Figure 5) [40]. This kind of games are classified as *sandbox* or *open-world*.

Figure 5. (**Left**) An example world constructed in Minecraft. Similar to Lego^{TM} blocks, no rule forces players to build anything specific. Creations come out of personal will, just because the game allows them. (**Right**) In Goat Simulator there is no specific reward for jumping over an ultralight but players do it because they can and it is fun: it is a way of experimenting, just like in the real world.

On latest years game designers are paying more attention to playfulness as sandbox or open world games are increasingly more demanded. Reasons for this demand have already been pointed out: players have complete autonomy for developing their own creativity in vast open decision spaces, they can pursue their own goals, experiment by trial and error, create their own personal challenges and constantly discover what happens as a result of their actions. Clearly, these kind of games properly address many of the items in this rubric, including playfulness ability. Some of these games have goals also but they let and encourage players to do anything they like, pursue goals in any order or even forget about goals and just explore and experiment. This is how these games become toys: they can be played in almost any imaginable way by players, similar to children playing invented games with a ball. The ball is a just a toy that can be used for any kind of play.

Playfulness is absent most of the time in educative environments. However, it is present on research and development environments. In fact, most of present discoveries in many branches of knowledge came out of experimental approaches. These approaches are completely similar to the playfulness nature of toys. Research has no intrinsic goals in general: it emerges from raw questions. Researchers are presented with current evidence and they ask questions about why or how things happen. That leads to experiments to seek answers and this gives new evidence. Evidence then gives ideas to developers to create. And all this cycle is driven by curiosity. Therefore, it could be considered a playfulness approach, as it is completely similar.

This characteristic is highly desirable in our educative environments. Curiosity is the most important driver for knowledge and a playfulness enabled environment fosters curiosity and experimentation. However, there is a great challenge involved. As teachers we usually design from knowledge to activities. This implies that activities are designed to practice and acquire some concrete knowledge or abilities. So activities tend to be the opposite of sandboxes: they are usually focused on some finite set of goals and give little or no space to trainees for experimentation or play. Great Gamification designs should change this focus and seek on producing sandbox-like activities.

2.10. Automation

One of the main differences in Figure 4 is due to automation. Characteristics like feedback, challenge or randomness are greatly affected by the level of automation. A game like Brain Age cannot exist without automation. Similar games can be made, even in manual contexts but they will be different to Brain Age.

Interaction based on the immediate feedback from a computer game generates great amounts of information per second. Players' brain subconsciously analyse cause/effect relations between this information and their input interactions. This fosters adaptations in players' brains as they advance practicing and mastering the game. This practical learning also happens on sports, which are real-life games outside a computer support.

When referring to a gamified activity, automation defines the level of human intervention required to produce responses to trainee's inputs. It also refers to the need of human intervention to enforce the rules. Computer games automatically process all inputs from players, give immediate responses and enforce the rules without any human intervention. By contrast, a group of players of a board game have to do all this processing: throwing dices, counting, interpreting rules, changing status of the game and so forth. Exactly the same happens with tests, exams or manual classroom activities.

Therefore, there are two relevant differences between manual and automated activities with respect to learning: the stream of information generated and the immediacy of the responses to input interactions. Both have been discussed in previous characteristics like feedback, challenge or learning by trial and error and have great impact on learning outcomes. For Gamification this means that automation should be sought always when possible. However, not all contexts are easy or viable to automate, nor every automation has to include computers. If we consider soccer, most of the game is automated. A referee is required to enforce the rules but most of its interactions are performed by the field, the ball, the goals and the players. In fact, a fan soccer match can be played without a referee. Same happens to other games and sports. This shows that some great level of automation can be achieved with appropriate real-life designs.

3. The Rubric

Table 1 shows the game-design-based rubric with the ten selected characteristics, their assessment criteria and assigned scores. The rubric has ten rows, one for each characteristic. Each row is divided into three columns that hold the criteria to assign scores from zero to two. For each characteristic, the given score will be at the top of the column that contains the criteria that more accurately describes the design. For simplicity, only a integer score is assignable to each characteristic.

The rubric has been designed as an instrument and so it meets the requirement to fit in a single page. To accomplish this, criteria have been written with a few simple words. This makes them simpler, more direct but less detailed and specific. It is advised to understand written criteria as a general contextual description. They are thought to be complemented with more detailed descriptions from Section 2. Also, criteria are written with three or four sentences per cell. For an appropriate application, they should not be considered a check-list: depending on the design being assessed they could even be not applicable as they are written. These sentences should better be considered as a description of general observable symptoms from designs that meet the criteria. This is a consequence of designs being artistic in nature: strict objective descriptions would not be applicable most of the time.

Table 1. Ten-characteristic game-design-based gamification rubric.

Characteristic	0	1	2
Open Decision Space	Not open No real decisions to take Only Correct/Incorrect	Decision-tree like Designed decision space With options but limited	Completely open Multiple/Infinite options Continuous decision spaces
Challenge	Single difficulty/activity No activity-ability match Punishments prevent beneficial attempts	Incremental difficulty Speculative Design Subjective matching Subjective measures	Sinusoidal difficulty progression Designed activity-ability match Measured, balanced, tested
Learning by Trial and Error	Failure punished Max.Marks only achievable without failure	Failure permitted Max.Marks achievable with some failures	Failure encouraged for learning Max.Marks achievable independent of failures
Progress Assessment	No progress measures No feedback on progress	Some progress measures defined Some feedback on status/progress Lack of precision	All progress defined All progress measured Detailed feedback on status/progress Next steps are clear
Feedback	None/minimal feedback response to actions Cause-effect learning is difficult/impossible	Some feedback response Some actions w/feedback Feedback not immediate Some cause-effect learning is possible	All actions produce cause-effect feedback Feedback immediate or timely adequate Cause-effect learning
Randomness	Everything is predictable No randomness involved No surprises	Some unpredictability Some random events or parts of activities Speculative/casual design of random parts	Measured unpredictable content and random parts of activities Purposively designed Surprises included, designed and balanced
Discovery	No new content No discovery No unlocking Content is fixed	Activities presents new content on progress Some unlockable content New content does not deliver surprises	New content is presented at a measured pace Discoverable content rewards user interest Surprises on discovery
Emotional Entailment	No design that targets emotions No characters, stories or aesthetics Focus on factual content	Some form of design to target emotions Use of template stories characters or aesthetics Imaginary experiences	Specifically-designed characters, stories and/or aesthetics Design focuses on creating an emotional experience
Playfulness Enabled	Concrete goals Specific procedures No room to experiment No curiosity generated	Selectable goals and/or procedures Room for development of personal creations Optional activities with creative component	Selectable/generable goals Creative procedures Users may play with goals, content and procedures in non-predesigned ways Curiosity rewarded
Automation	No automation Manual intervention All or most of the rules are manually enforced Slow feedback response time	Some level of automation Optimized manual intervention Rules are partly enforced on an automatic way Improved feedback response time	Everything automated None or minimal manual intervention required Rules are/can be enforced automatically Immediate or fastest feedback response time

Two main outcomes arise from the use of the rubric as proposed: first, the rubric is an easy to use instrument for assessing strengths and weaknesses of designs with respect to their game-like characteristics. Second, the knowledge of strengths and weaknesses helps thinking in ways to improve designs, creating a feedback cycle of analysis and improvement. These goodnesses are limited by the subjective nature of the rubric and the discretization it imposes over the analysis space to only ten characteristics and three scores. However, these limitations are acceptable and even desirable in the selected context of helping new practitioners to overcome the lack-of-experience entry barrier.

4. Rubric Application Samples

As an initial piece of evidence, we show four samples of application of the rubric to four different activities in two blocks: Sections 4.1 and 4.2 analyse the *Super Mario Bros* game and a unsuccessfully gamified 16-week course described in literature. Sections 4.3 and 4.4 analyse the activity of solving a single system of linear equations and a gamified activity designed based on linear equations systems solving. The first pair of examples shows how the rubric compares a successful game with an unsuccessful gamification expecting a great difference in their scores. The second pair shows how the rubric measures the difference between a single classic learning activity and a gamification design produced with the items of the rubric in mind. This gives an idea on how the rubric could be used to help practitioners create and improve their initial designs.

This small piece of evidence does not prove the general validity of the rubric but yields a initial hint. More support evidence is required in any case to validate or discard the rubric.

4.1. Super Mario Bros (NES)

Next we will assess the game *Super Mario Bros* [30] (see Figure 1). The player controls Mario, a plumber in an imaginary world that has to save his princess from an enemy. For that, Mario has to surpass many perils and enemies in a series of levels. Mario's abilities include running and jumping, getting inside pipes, breaking blocks with a punch from below and firing. *Super Mario Bros* is classified as an action-platformer game: most of the time Mario has to jump from platform to platform to surpass the perils.

Super Mario Bros is considered one of the most played games in videogame history. Millions of people have played either the original game or any of its successors. Let us apply the rubric and confirm if this popularity correlates with its score:

- [2] *Open Decision Space*. The game lets the user take movement decisions (actions) in a continuous world. Taking any two players that successfully finish one level, it is almost impossible that both of them perform the exact same actions. Player is in total control of the action having potentially infinite options in a continuous space.
- [2] *Challenge*. The game is composed of a series well designed levels to challenge players. Difficulty progression is sinusoidal, with some easier levels after more challenging ones. It is balanced and tested by designer intuition through iterations.
- [2] *Learning by trial and error*. As many games, learning by trial and error is in the very core of the game. Failure is permitted with a number of lives in one game but there is no limit of games per player. The player can complete the game regardless of the number of games or lives lost to learn. Even level design is thought to encourage players to learn by experimenting.
- [1] *Progress assessment*. The game assesses the progress of the player through levels and player status. Whenever a level is finished, the player does not repeat it even if lives are lost. Inside a level, the player always knows how to continue to achieve the end and feedback through movement, points, enemies and music reports the progress.
- [2] *Feedback*. Similar to most action-platformer games, there are sixty frames per second of continuous cause/effect feedback that lets the player sense control and learn. Moreover, game design informs of all events happening such us lives lost, enemies beaten, objects obtained and so forth.

- [1] *Randomness.* Although the game is predictable, with no actual random events happening, there are some enemies with elaborated movements that give the player some sense of unpredictability.
- [2] *Discovery.* Players discover new levels and worlds as they finish previous ones, in an unlock-like fashion. There are secret places, items and bonuses at different locations that reward players for their attention to detail and exploration. Also, there are some special behaviours of game elements that can be discovered by experimentation.
- [2] *Emotional entailment.* The complete game creates an emotional experience for the player with the aesthetics, characters, music and the story. It is completely conceived as an adventure in an imaginary world where characters live and become "real" in some sense for the player.
- [1] *Playfulness enabled.* Although the game has clear goals and rules, players have room to explore and be creative. In fact, communities of players have engaged in new challenges like the *speed-run* modalities, creating new rules on top of the game. The game was not thought to be played as a toy but players can and use it this way.
- [2] *Automation.* As a console game, meant to be played at home, the game is fully automated. Feedback is immediate and all rules are enforced automatically.

According to the rubric, *Super Mario Bros* has an score of $2+2+2+1+2+1+2+2+1+2 = 17$ points, which is quite reasonable for such a well known and played game.

4.2. Unsuccessful 16-Week Gamified Semester

As a more elaborated example we will apply the rubric to the Gamification study by D. Hanus et al. [41]. A class was divided into two groups. The control group received normal lessons, materials, assignments and exams. The experimental group was given same content than the control group but in a standard gamified fashion including badges, leaderboards and incentive systems. Badges were given as a reward for positive behaviours like interaction with class materials, study in pairs in the library or handing assignments early. There also was a badge for entering lessons dressed up like a videogame character. In addition to badges, students also earned coins for small contributions to class discussions or sharing interesting information. Students could use coins to earn some class benefits like extension on a paper.

Students were required to obtain some mandatory badges but coins were optional. The leaderboard was ordered by number of badges obtained, with students using pseudonyms. The leaderboard was updated weekly.

The description of the system is too broad, which produces a great level of noise. Therefore, the final score from the rubric should be taken with care. Adding a big error bar to the final result is advisable to compensate the noise. In raw application, the rubric yields these scores:

- [1] *Open Decision Space.* Although badges are reported as mandatory, some seem to be optional like coins. Also students can decide how to use earned coins. However, this seems like a small set of options with not much strategy involved.
- [0] *Challenge.* No description of badges seems to match with levels of difficulty or abilities required. They are focused on behaviour. There seems to be no consideration about difficulty.
- [0] *Learning by trial and error.* No consideration of opportunities, number of assignments or punishments.
- [1] *Progress assessment.* Number of badges earned, coins and the leaderboard give some form of progress assessment.
- [1] *Feedback.* Feedback seems to come mainly from teacher and the leaderboard is updated once a week. Therefore, feedback seems rather slow.
- [0] *Randomness.* Everything described in the system seems concretely specified, with no room for surprises or random events.
- [0] *Discovery.* Similarly, as everything seems predefined, there is nothing to unlock or discover.

- [0] *Emotional entailment.* There are no characters, no story, no aesthetics. The only content that could be related to emotions is the badge for dressing up like a videogame character.
- [0] *Playfulness enabled.* Similarly, description of the system does not involve any ability to use activities as toys or even play with strategies. Some creativity could be exhibited with the videogame character dressing badge or in the way to expend coins.
- [0] *Automation.* Nothing is automated. Even badges have to be claimed by students by filling up forms. However, the leaderboard being updated weekly can be perceived as some small form of automation by students.

The rubric gives a final score of 3 points for this Gamification design. Even considering an important error bar up to 100%, maximum value would be 6 points, really far from the 17 points obtained by *Super Mario Bros*. It clearly appears not to be enough to induce important motivational changes on students.

This analysis supports results obtained by D. Hanus et al. [41], who concluded that the Gamification methods they used had no positive impact on learning and could even harm student motivation. As the 3 points obtained are far lower compared to *Super Mario Bros*, a much inferior motivational level could be expected. This result is supportive of what D. Hanus et al. found in their study.

4.3. Single Learning Activity: Solving a Linear-Equations System

To establish a comparison with a common learning activity, we will now apply the rubric to a linear-equations system exercise. Let us consider a trainee solving the system on paper and handing it to the teacher. A week afterwards, the trainee receives the exercise assessed. These would be the rubric scores:

- [1] *Open Decision Space.* Some minimal decisions can be considered regarding the solution method and the order in which to perform steps.
- [0] *Challenge.* It is a single activity, so there is no way to match activity with ability. Difficulty is fixed.
- [0] *Learning by trial and error.* While the trainee produces its solution there is no feedback, no way to know if decisions are good or bad. Therefore, no way to cause/effect learn.
- [0] *Progress assessment.* The only perceivable progress would be the steps done towards the solution but that is no form of progress assessment.
- [0] *Feedback.* There is no feedback response to actions and teacher feedback takes one week. Cause/effect learning is almost impossible.
- [0] *Randomness.* Everything is completely predictable and there are no surprises.
- [0] *Discovery.* As all content is fixed, there is no unlocking or discovery at all.
- [0] *Emotional entailment.* There are no characters, no story, no aesthetics, no content that could be related to emotions.
- [1] *Playfulness enabled.* There is some room to experiment with procedures or methods, but very limited.
- [0] *Automation.* Everything is manually performed, with no automation at all and a slow response time (one week).

This gives 2 points of final score for the learning activity in isolation. It clearly contrasts with *Super Mario Bros* and shows a strong difference. Similar strong difference is usually perceived on student motivation on these two activities. Both scores seem intuitively correlated with this general perception.

4.4. Gamified Version of the Linear-Equations Solving Activity

Now we consider a explicit gamification design created for the activity of solving linear-equations. This activity was presented by Llorens et al. [42]: here we present a summary of the design along with its evaluation using the rubric. For complete details on the design please refer to Reference [42]. Basically, Llorens et al. propose these changes to the activity:

1. Create an automatic generator of linear-equation systems to present students with hundreds of exercises instead of one.
2. Classify generated systems into 6 levels of difficulty depending on their intrinsic characteristics, number of variables and numerical complexity.
3. Form teams during solving sessions and have rules to require teams and individuals to develop strategies to distribute tasks and face challenges.
4. Give points to valid solutions depending on the assigned difficulty of the system.
5. Make difficulty levels unlockable and have clear unlocking rules to force them to appropriately master levels before proceeding.
6. Have a student experience level (XP) that increases as students solve systems and successfully resolve proposed activities. Define experience levels and use them as a measure to form teams and unlock difficulty levels.
7. Spread the activity across many sessions and maintain points, experience and levels. Let students evolve over the course.
8. Produce random events that interrupt sessions and change rules surprisingly. Examples: A fleeting system that has to be solved fast, a red-code event in which students have to deactivate a bomb or a dizzy time during which solutions have to be given inverted.
9. Define a set of achievements to give to students, including some secret ones to reward their research or detailed abilities.
10. Automatize all the system with an application that lets students select systems, send solutions and receive instant status reports with their mobile phones.

Let us now compare the evaluation of this gamified version to the single linear-equations system solving activity:

- [1] *Open Decision Space*. Students have the freedom to define different strategies to solve tasks and challenges based on linear-equations systems.
- [1] *Challenge*. There are different difficulty levels defined as progressive and linear.
- [2] *Learning by trial and error*. Students are limited by factors like time during sessions but not by their mistakes. They can fail many times and continue, not limiting their final score.
- [2] *Progress assessment*. There are several measures like experience points, regular points, levels, achievements and unlocked difficulties that give students great detail on their progress.
- [1] *Feedback*. Students receive feedback from the system with respect to their solutions and actions. They see their progress and know if they have done right or wrong and they also can fix their failures. Feedback is not complete and sometimes cause-effect relationships maybe diffuse.
- [2] *Randomness*. Systems are generated, so randomness is present most of the time in the system. Moreover, random events are another source of purposively designed unpredictability for students, which induces surprises.
- [1] *Discovery*. There is some unlockable content and some secret levels and achievements, but design could be improved to include more surprises and learning through discovery.
- [1] *Emotional entailment*. Random events are based on simple stories like deactivating a bomb, for instance. Also, time limitations and surprises target emotions but there is a lack of a general story, some characters and appropriate aesthetics.
- [0] *Playfulness enabled*. There is a small subset of creativity involved in the way teams can approach tasks, but goals are clearly defined and there is not much room for free playing outside the rules of the system.
- [2] *Automation*. A mobile app with a server give a moderate level of automation and control with minimal manual intervention required. Feedback is immediate to responses, although it

might be not detailed or complete. However, this last part is improbable with new versions of the app.

This gamified version of the activity gets a score of $1+1+2+2+1+2+1+1+0+2 = 13$ points in total. Improvement over the 2 points obtained by the single activity is clear, even considering possible variable criteria interpreting the rubric that could lead to a reduction of some points. In this sense, the rubric also shows potential for helping practitioners to improve their designs: the proposed design could have been made by looking at the items of the rubric and including ideas to improve each one of them. An experience designer would probably consider the design as a whole product instead of a separate set of characteristics. However, the set of separate characteristics is much more easily manageable for a trainee and can easily lead to initial designs like the one proposed here.

5. Conclusions

In this paper we have presented a rubric as an instrument to help new Gamification practitioners assessing designs. Its aim is lowering the entry barrier to get experience in game-design-based Gamification. Experience is obtained through practice but practice without guidance is much more difficult and frustrating. The rubric gives this guidance.

The rubric can be used to assess a given design, to analyse strengths and weaknesses and to highlight areas for improvement. This uses are focused on helping practitioners learn and develop experience on game-design-based Gamification.

Due to the artistic nature of Game Design, the rubric is conceived on previous experience from authors and works from experts. The rubric itself is conceived with flexible and interpretable criteria to fit all subtle perceptional details from games and Gamification experiences. This is both a limitation and a strength: it cannot provide objective assessment but it allows considering emotions and player experiences, which are key for a successful Gamification design.

As practical experience is the main basis for successful game-design-based Gamification designs, the rubric is probably be far too simplified for experienced practitioners. This is an intended limitation, as its focus is to help new practitioners.

Four samples of application of the rubric have been shown in two pairs: to *Super Mario Bros* game and a 16-week gamified course presented by D. Hanus et al. [41] and to a linear-equations solving exercise and a gamified version of the linear-equations solving exercise proposed by Llorens et al. [42]. The four applications yield consistent results with previous evidence and general perception. Although much more evidence is required to assess the general validity of the rubric, this piece of evidence encourages further testing and analysis.

Author Contributions: Conceptualization, F.J.G.-D., C.J.V.-A., R.M.-C. and F.L.-L.; Investigation, R.S.-C. and P.C.-R.; Writing—original draft, F.J.G.-D. and C.J.V.-A.; Writing—review & editing, F.J.G.-D., C.J.V.-A., R.M.-C. and F.L.-L.

Funding: This research received no external funding

Conflicts of Interest: The authors declare no conflict of interest.

References

1. Deterding, S.; Dixon, D.; Khaled, R.; Nacke, L. From Game Design Elements to Gamefulness: Defining "Gamification". In Proceedings of the 15th International Academic MindTrek Conference: Envisioning Future Media Environments, Tampere, Finland, 28–30 September 2011; ACM: New York, NY, USA, 2011; pp. 9–15. [CrossRef]
2. Nacke, L.E.; Deterding, S. The maturing of gamification research. *Comput. Hum. Behav.* **2017**, *71*, 450–454. [CrossRef]
3. Linehan, C.; Kirman, B.; Lawson, S.; Chan, G. Practical, Appropriate, Empirically-validated Guidelines for Designing Educational Games. In Proceedings of the SIGCHI Conference on Human Factors in Computing Systems, Vancouver, BC, Canada, 7–12 May 2011; ACM: New York, NY, USA, 2011; pp. 1979–1988. [CrossRef]

4. Nah, F.F.H.; Zeng, Q.; Telaprolu, V.R.; Ayyappa, A.P.; Eschenbrenner, B. Gamification of Education: A Review of Literature. In *Lecture Notes in Computer Science*; Nah, F.F.H., Ed.; Springer International Publishing: Cham, Switzerland, 2014; pp. 401–409. [CrossRef]
5. Hamari, J.; Koivisto, J.; Sarsa, H. Does Gamification Work?—A Literature Review of Empirical Studies on Gamification. In Proceedings of the 47th Hawaii International Conference on System Sciences, Waikoloa, HI, USA, 6–9 January 2014. [CrossRef]
6. Fitz-Walter, Z.; Johnson, D.; Wyeth, P.; Tjondronegoro, D.; Scott-Parker, B. Driven to drive? Investigating the effect of gamification on learner driver behavior, perceived motivation and user experience. *Comput. Hum. Behav.* **2017**, *71*, 586–595. [CrossRef]
7. Kocakoyun, S.; Ozdamli, F. A Review of Research on Gamification Approach in Education. In *Socialization*; Morese, R., Palermo, S., Nervo, J., Eds.; IntechOpen: Rijeka, Croatia, 2018; Chapter 4. [CrossRef]
8. Koivisto, J.; Hamari, J. The rise of motivational information systems: A review of gamification research. *Int. J. Inf. Manag.* **2019**, *45*, 191–210. [CrossRef]
9. Werbach, K. (Re)Defining Gamification: A Process Approach. In *Persuasive Technology*; Spagnolli, A., Chittaro, L., Gamberini, L., Eds.; Springer International Publishing: Cham, Switzerland, 2014; pp. 266–272. [CrossRef]
10. Chou, Y. *Actionable Gamification: Beyond Points, Badges, and Leaderboards*; Createspace Independent Publishing Platform: Milpitas, CA, USA, 2015.
11. Deterding, S. Eudaimonic Design, or: Six Invitations to Rethink Gamification. In *Rethinking Gamification*; Fuchs, M., Fizek, S., Ruffino, P., Schrape, N., Eds.; Meson Press: Luneburg, Germany, 2014; pp. 305–331
12. Koster, R. *A Theory of Fun for Game Design*; Paraglyph Press: Scottsdale, AZ, USA, 2004.
13. Schell, J. *The Art of Game Design: A Book of Lenses*; Morgan Kaufmann Publishers Inc.: San Francisco, CA, USA, 2008.
14. Gee, J.P. *What Video Games Have to Teach Us About Learning and Literacy. Second Edition: Revised and Updated Edition*; Palgrave Macmillan: New York, NY, USA, 2007. OCLC: ocn172569526.
15. Gee, J.P. *Good Video Games and Good Learning*; Peter Lang Inc., International Academic Publishers: New York, NY, USA, 2007.
16. Kreimeier, B. The Case For Game Design Patterns; Gamasutra Featured Article; Gamasutra, 2002. Available online: https://www.gamasutra.com/view/feature/132649/the_case_for_game_design_patterns.php (accessed on 4 November 2019)
17. Lindley, C.A. Game Taxonomies: A High Level Framework for Game Analysis and Design; Gamasutra Featured Article; Gamasutra, 2003. Available online: https://www.gamasutra.com/view/feature/131205/game_taxonomies_a_high_level_.php (accessed on 4 November 2019)
18. Hunicke, R.; Leblanc, M.; Zubek, R. *MDA: A Formal Approach to Game Design and Game Research*; AAAI Workshop: Challentes In Game Artificial Intelligence, Volume 1; Technical Report; AAAI: Menlo Park, CA, USA, 2004.
19. Reeves, B.; Read, L. *Total Engagement: Using Games and Virtual Worlds to Change the Way People Work and Businesses Compete*; Harvard Business Review Press: Boston, MA, USA, 2009.
20. Tondello, G.F.; Kappen, D.L.; Mekler, E.D.; Ganaba, M.; Nacke, L.E. Heuristic Evaluation for Gameful Design. In Proceedings of the 2016 Annual Symposium on Computer-Human Interaction in Play Companion Extended Abstracts, Austin, TX, USA, 16–19 October 2016 ; ACM: New York, NY, USA, 2016; pp. 315–323. [CrossRef]
21. Rapp, A. Designing interactive systems through a game lens: An ethnographic approach. *Comput. Hum. Behav.* **2017**, *71*, 455–468. [CrossRef]
22. Strawson, P.F.; Wittgenstein, L. Philosophical Investigations. *Mind* **1954**, *63*, 70. [CrossRef]
23. Bogost, I. Persuasive Games: Exploitationware. Gamasutra, 2011. Available online: http://www.gamasutra.com/view/feature/6366/persuasive_games_exploitationware.php (accessed on 29 June 2019).
24. Desurvire, H.; Wiberg, C. *Game Usability Heuristics (PLAY) for Evaluating and Designing Better Games: The Next Iteration*; Online Communities and Social Computing; Springer: Berlin/Heidelberg, Germany, 2009; pp. 557–566. [CrossRef]
25. Tromp, N.; Hekkert, P.; Verbeek, P.P. Design for Socially Responsible Behavior: A Classification of Influence Based on Intended User Experience. *Des. Issues* **2011**, *27*, 3–19. [CrossRef]

26. Linehan, C.; Bellord, G.; Kirman, B.; Morford, Z.H.; Roche, B. Learning Curves: Analysing Pace and Challenge in Four Successful Puzzle Games. In Proceedings of the First ACM SIGCHI Annual Symposium On Computer-Human Interaction in Play—CHI PLAY, Toronto, ON, Canada, 19–21 October 2014; ACM Press: New York, NY, USA, 2014; pp. 181–190. [CrossRef]
27. Llorens-Largo, F.; Gallego-Durán, F.J.; Villagrá-Arnedo, C.J.; Compañ Rosique, P.; Satorre-Cuerda, R.; Molina-Carmona, R. Gamification of the Learning Process: Lessons Learned. *IEEE Rev. Iberoamericana Tecnol. Aprendizaje* **2016**, *11*, 227–234. [CrossRef]
28. Villagrá-Arnedo, C.; Gallego-Durán, F.J.; Molina-Carmona, R.; Llorens-Largo, F. PLMan: Towards a Gamified Learning System. In *Learning and Collaboration Technologies*; Zaphiris, P., Ioannou, A., Eds.; Springer International Publishing: Cham, Switzerland, 2016; Volume 9753, pp. 82–93.
29. Deci, E.L.; Ryan, R.M. *Handbook of Self-Determination Research*; University Rochester Press: Rochester, NY, USA, 2004.
30. Demaine, E.D.; Grandoni, F. (Eds.) *Super Mario Bros. is Harder/Easier Than We Thought, Leibniz International Proceedings in Informatics (LIPIcs)*; Schloss Dagstuhl–Leibniz-Zentrum fuer Informatik: Dagstuhl, Germany, 2016; Volume 49. [CrossRef]
31. Nacke, L.; Lindley, C.A. Flow and Immersion in First-person Shooters: Measuring the Player's Gameplay Experience. In Proceedings of the 2008 Conference on Future Play: Research, Play, Share, Toronto, ON, Canada, 3–5 November 2008; ACM: New York, NY, USA, 2008; pp. 81–88. [CrossRef]
32. Csikszentmihalyi, M. *Flow: The Psychology of Optimal Experience*; Harper Perennial: New York, NY, USA, 1991.
33. Selye, H. The general adaptation syndrome and the diseases of adaptation. *J. Allergy Clin. Immunol.* **1946**, *17*, 231–247. [CrossRef]
34. Forget, A.; Chiasson, S.; Biddle, R. Lessons from Brain Age on persuasion for computer security. In Proceedings of the 27th International Conference on Human Factors in Computing Systems, Boston, MA, USA, 4–9 April 2009; pp. 4435–4440. [CrossRef]
35. Ayton, P.; Fischer, I. The Hot Hand Fallacy and the Gambler's Fallacy: Two faces of Subjective Randomness? *Mem. Cognit.* **2005**, *32*, 1369–1378. [CrossRef] [PubMed]
36. Shannon, C.E. A mathematical theory of communication. *Bell Syst. Tech. J.* **1948**, *27*, 379–423. [CrossRef]
37. Prensky, M. *Digital Game-Based Learning*; McGraw-Hill: New York, NY, USA, 2001.
38. Prensky, M. *Don't Bother Me Mom–I'M Learning!* Paragon House Publishers: St.Paul, MN, USA, 2006.
39. Huizinga, J. *Homo Ludens: A Study of the Play-Element in Culture*; Beacon Press: Boston, MA, USA, 1955.
40. Jensen, L.J.; Barreto, D.; Valentine, K.D. Toward Broader Definitions of "Video Games": Shifts in Narrative, Player Goals, Subject Matter, and Digital Play Environments. In *Examining the Evolution of Gaming and Its Impact on Social, Cultural, and Political Perspectives*; IGI Global: Hershey, PA, USA, 2016; pp. 1–37. [CrossRef]
41. Hanus, M.D.; Fox, J. Assessing the effects of gamification in the classroom: A longitudinal study on intrinsic motivation, social comparison, satisfaction, effort, and academic performance. *Comput. Educ.* **2015**, *80*, 152–161. [CrossRef]
42. Llorens-Largo, F.; Molina-Carmona, R.; Gallego-Durán, F.J.; Villagrá-Arnedo, C.J. Guía para la Gamificación de Actividades de Aprendizaje. *Novatica* **2018**. Available online: https://www.novatica.es/guia-para-la-gamificacion-de-actividades-de-aprendizaje (accessed on 4 November 2019).

© 2019 by the authors. Licensee MDPI, Basel, Switzerland. This article is an open access article distributed under the terms and conditions of the Creative Commons Attribution (CC BY) license (http://creativecommons.org/licenses/by/4.0/).

Article

Serious Games, Mental Images, and Participatory Mapping: Reflections on a Set of Enabling Tools for Capacity Building

Teodora Iulia Constantinescu * and Oswald Devisch

Architecture Department—Spatial Capacity Building Group, Hasselt University, B-3590 Diepenbeek, Belgium; oswald.devisch@uhasselt.be
* Correspondence: teodora.constantinescu@uhasselt.be

Received: 29 January 2020; Accepted: 25 February 2020; Published: 3 March 2020

Abstract: Increasing complexity of societal questions requires participatory processes that engage with capable participants. We adopted Horellis' stance on participation as not an isolated event but a constant communication between different groups that can be assured by using enabling tools. We applied the Capability Approach to frame a capacity-building process and understand how this framework can support a collective of entrepreneurs to become aware of their capabilities (and the impact of an ongoing urban renewal process on these capabilities). The Capability Approach emphasizes the personal and structural conditions that impact a person's capability to choose—the conditions that affect the process of determining what a person values. The paper builds on a two year capacity-building process conducted in Genk, Belgium, and proposes a conceptual framework for building capacities, in which the process and outputs collide with ideas of *choice, ability, and opportunity*, notions central to the Capability Approach. The case study looks at one of the main commercial streets of the city (Vennestraat) and reflects on a set of enabling artefacts used to engage proprietors in the capacity-building process. This capacity-building process, characterized by the idea of space and capabilities, advances a critical viewpoint on issues related to participatory processes and gives practitioners a set of enabling tools to start a conversation over complex urban transformations, such as the one in Vennestraat.

Keywords: capability approach; capacity building; serious games; enabling tools; mental images

1. Introduction

Societal processes are growing increasingly more complex (e.g., climate change is not only an environmental issue but also a social one, and design for urban spaces that foster these processes is becoming more and more people-oriented [1]. Handling these complex contexts may mean conceptualizing design "not as the creation of discrete, intrinsically meaningful objects, but the cultural production of new forms of practice" [2]). This translates into the need to understand the mechanisms of 'meaning making' in order to support and improve building capacities [2]. This article discusses three challenges: (1) the reality that some groups lack the opportunity to engage in such complex design processes, (2) the growing need to strengthen their capabilities to participate in the said processes, and (3) employing tools that can give form to citizens' capabilities in said processes.

We address these challenges through the Vennestraat case—a main commercial street in Genk, BE, shaped by various waves of migrant entrepreneurs that had to respond to the ever-changing economic dynamics of the city. We discuss a project where researchers together with citizens underwent a two-year capacity-building process via a set of enabling tools to support communicative transactions that we refer to as enabling tools/artefacts due to their potential to develop participants skills and involve them into the debate on the refurbishment process of the street. We used Sen's and Nussbaum's Capability Approach to underline the dynamics between the different artefacts and understand their impact when used to fuel our capacity-building process. The concept of enabling tools is based on Horellis' [3] stance on participation as not an isolated event, but a constant communication between different groups that can be assured by using different methods.

Sen's and Nussbaum's Capability Approach advances a framework for 'planning, monitoring, and evaluation of development programs' [4] and moves past ones' capacities and aptitudes by getting a handle on peoples' chances and decisions to accomplish the things they value—their capability. 'This is where the main distinction between capacity and capabilities lies' [4]. The Capability Approach assumes that, without suitable chances, capacity building programs meant to advance new aptitudes and skills, will not suffice to empower individuals to accomplish that of most importance to them. As such, a capacity-building process should be a non-linear and essential process, grounded in understanding space as neither a medium nor an inventory of ingredients, but a mixture of geographies, built environment, symbolism, life practices and opportunities, interlinking each other. Through the Capability Approach lens, a capacity-building process: (1) facilitates practices that enable 'new forms of collective enunciations, making sense of the sensible' [5] and (2) transforms the design process into a set of methods and artefacts [6,7].

This article explores the following questions: (i) (how) Can we build capabilities of diverse actors to see and make sense of the systems they are part of? and (ii) What are the advantages of using the Capability Approach in a capacity-building process? We answer these questions as follows: in the first section, we discuss the Capability Approach—its mechanisms and components; in the second section, we analyze how the capacity-building process took shape through participatory methodologies. In the third section, we detail on which enabling tools were used. We conclude with how these artefacts supported the process of building capabilities.

2. The Promise of the Capability Approach Framework

Trying to assess social change in the economic context of the 1980s, Amartya Sen and Martha Nussbaum set up the premise of a new theory—the Capability Approach. The conceptual roots of the approach trace back to Adam Smith and Karl Marx, all the way to Aristotle's' use of 'life in the sense of activity' [8]. The Capability Approach (also dubbed the capabilities approach) centers around different ideas on what people 'are able to do (i.e., capable of)' [8]. It assesses societal transformation in terms of 'what are people actually able to do and to be' [8–10] in the framework of available policies—'opportunities people have when, and only when, policy choices put them in a position to function effectively' [9]. By definition, the framework concentrates on what people can do, that is, their ability.

A central concept in the Capability Approach are 'functionings' "what a person manages to do or to be" [11]. Complementarily, Alkire [12] claims that functioning/s is a general term for the activities, resources, and attitudes that people consider as significant. They may include dignity, an educated mind, knowledge, and warm friendship, among others. Capabilities, on the other hand, are liberties that help people to achieve these functionings. Noteworthy, a functioning refers to achievement, while the capability to a set of opportunities [13]. The approach is people-centered, meaning it dwells on what they value and aspire; its principal goal is to create an enabling environment for people to achieve what they value. While skills and abilities are crucial, the framework additionally aims at grasping choices and opportunities to facilitate individuals to achieve their aspirations [13]. Whereas the functioning relates to improving the abilities of individuals, capabilities, as a concept, it involves the idea of increasing the choices available, as well as preparing people to grasp opportunities. The substantial assumption of the Capability Approach is that, in the absence of proper opportunities and capacity building programs, the environment is not entirely sufficient to enable people to achieve their goals.

Based on Sen's [14] writings, the assessment of social arrangements is not entirely concerned with choice. They are also concerned with the different aspects that convert capabilities into realized functionings [15]. It also emphasizes the significance of personal and structural conditions that impact a person's capability to choose. These conditions function as conversion elements as they affect how choices become achievements (see Figure 1). In particular, *three conversion factors* influence ones' freedoms: (1) personal features (including intelligence, metabolism, and sex) determine the choices made by an individual; (2) social characteristics have pronounced contributions—they include social norms, hierarchical structures, and public policies; and (3) environmental aspects, such as public goods, climate, and infrastructure [15] (see Figure 1). Intrinsically, freedom relates to the amount of choices one has and the actual opportunities to perform the said choices. The Capability Approach emphasizes the personal and structural conditions that impact a person's capability to choose—the conditions that affect the process of determining what a person values (see Figure 1). As such, it becomes vital to consider the characteristics to determine the propensity for bringing change rooted in individual decisions.

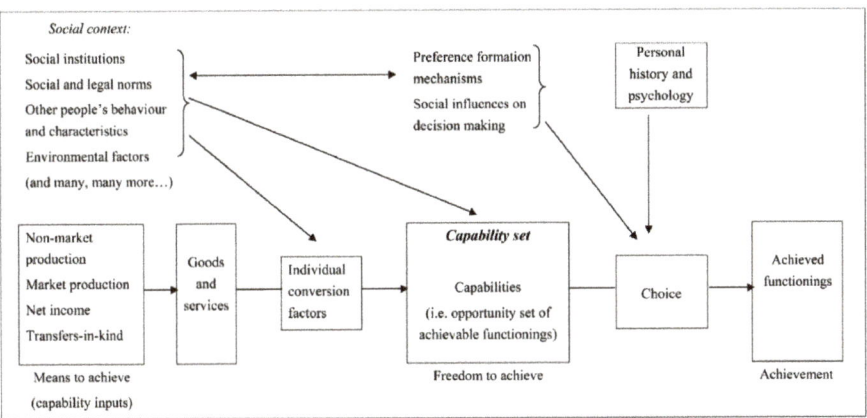

Figure 1. A stylized non-dynamic representation of a person's capability set and her social personal context [15].

'First, personal conversion factors (e.g., metabolism, physical condition, sex, reading skills, intelligence) influence how a person can convert the characteristics of the commodity into a functioning. Second, social conversion factors (e.g., public policies, social norms, discriminating practices, gender roles, societal hierarchies, power relations) and, third, environmental conversion factors (e.g., climate, geographical location) play a role in the conversion from characteristics of the good to the individual functioning' [15]. Thus, knowing the goods one possesses or can utilize does not suffice to understand which functionings he/she can accomplish; we must discover more about the individual and the conditions in which he/she lives. Citizen's freedom and their capability in the design process stems from their abilities, choices, opportunities and results [6,7]. The functionings—'the doings and beings valued by citizens', as defined by Sen (2005), can be a result of processes shaping deliberation, participation, and engagement [16]. A notable feature of this element is that its objective is to unpack the underlying values related to building capacities (i.e., in order to increase the freedom one has, we need to focus on the aforementioned conversion factors). Capacity building increases the options individuals have, and it guides the elaboration of techniques for deliberation, which reveals, builds, and addresses the conversion factors impacting the transformation of tactics into an achieved participatory principle. Based on the Capability Approach, we outline three specific elements of freedom—*choice, ability, and opportunity*—essential to a capacity building process. **Choice** is a critical element in a capacity building process: the provision of alternatives and increasing ones' possibilities is essential to improve democracy among citizens and enables them to participate actively. **Ability** refers to (1) how individuals engage in the deliberation process and (2) their relationship with different actors involved in this process (e.g., professionals-architects, engineers; policymakers). It equally relates to their capacities to appropriate diverse design decisions, like urban interventions. ***Opportunity*** stems from social, political, and economic processes that shape citizens freedoms. It requires engagement in the deliberation phase of the capacity building process. In process freedom, it is necessary to investigate ones' stance to understand if they have assumed the role of partners, informants, or actors of change in the process to assess the effectiveness of participation. Additionally, implementing a project in a specific area is the surest way of determining the propensity for expansion or a change in policymaking.

We propose a conceptual framework for building capacities by using three enabling tools (i.e., participatory mapping, mental images and serious games), in a process structured around the main ideas of freedom—choice, ability, and opportunity—notions central to the Capability Approach. The literature review exposes the need for understanding the conversion factors an individual has when designing for "a dynamic multiplicity, in which city-making envisages and is organized as an inclusive and responsive process" [17]. This capacity-building framework, framed by the idea of space and capabilities gives a set of enabling tools to practitioners that deal with complex urban transformation projects at different scales.

In order to increase the freedom of a person, he/she needs to get an understanding of his/hers conversion factors (and the elements that frame it). As individuals are prone to suffer from system blindness [18], we embark in the endeavor of assisting them to make sense of the systems they are part of. Given the complexity of this system, the process of 'making things visible' needs to be a collective process—a process during which a variety of people discuss (1) the elements that block or support them in achieving their needs and aspirations and (2) each others' roles in the said process. In this paper, we explore how we can support this collective process, based on Horellis' [3] stance, and that it requires enabling tools/artefacts. As such, we use playful environments (i.e., participatory mapping), gamified enquiries (i.e., mental images), and serious games as enabling tools to (1) support conversations between different (groups of) actors about complex topics, such as urban renewal projects, and (2) uncover the tacit knowledge and shared learning embedded in local networks, critical to trace and understand ecosystem dynamics at urban scale [19,20]. We define enabling tools by linking them with the concept of participation as described by Pelle Ehn [21]: artefacts/methods that create meeting points between language games of people with each their expertise. Enabling tools/artefacts are 'formatted spaces of participation' [22] in that they are technologically, socially, and economically pre-structured interfaces via which citizens can perform certain actions. That implies that, while the experience of capacity-building processes can turn out to be progressively available for the difficult to-arrive at groups and can assist members with articulating their dreams and wants, it will consistently be in the arrangement directed by the specific artefact used in that situation. The strain between the given structure of the capacity-building process and its steady redefinition by the act of participation will be constant. We use games as structured enabling tools, complemented by very open ones (i.e., mental images, participatory mapping). We need such tools to support people in the endeavour of understanding their conversion factors and, thus, their freedoms.

The contribution we are aiming at is operational and theoretical: we situate a capacity-building process as a design part of a broader participatory process. In what follows, we present and unpack our reasoning: we first look into the context and how we have built our reasoning when designing the capacity-building process. We do so through different enabling tools (e.g., serious games, participatory mapping, mapping mental images, and interviews) used to (1) learn about the social, political and economic reality of the context, (2) support people to make sense of the systems they are part of, and (3) understand the potential and limitations of each artefact. Finally, we tie everything together by framing the capacity-building process via a set of effective and consistent participatory methodologies.

3. The Context

Genk's development has always been related to work intensive industrial activities [23] relying on mining, until the 1980s and, later on, on car manufacturing, namely Ford Automotive. However, the closing of the coal mines in the mid-1980s, followed by the closure of Ford industries

(2014), tempered the city's economic development. As a reaction to the two main economic shifts, the closure of both mining and car industries, the city of Genk decided to change its development strategy and invested in (1) cultural-, research-, and art-related infrastructures and (2) developing circular economy principles and logistics. The city drafted a new project for the former coal mine and transformed it into a cultural hub, C-mine, where C stands for creativity. C-mine was established on the old mining site reviving the existing buildings. The municipality has owned the complex since 2001. As a second measure, to combat the consequences of the closure of the car factory in 2014, the government drafted the SALK plan (Strategic Action Plan) to make the city competitive again. The plan describes a series of measures designed to provide 3000 jobs in the short-term and 10,000 jobs in the long-term, with an implementation period from 2013 to 2019.

Vennestraat, a well-known shopping street, is one of Genk's main commercial streets which met its significant growth at the time of the coal mine. After the mine closed, the street started to decay. It was shaped by different waves of entrepreneurs over the years, starting with a majority of Turks, followed by Italians, making way to a multicultural mix of proprietors today. Fueled by the arrival of C-mine in its immediate vicinity, Vennestraat experienced gradual upgrades in the past decade. C-mine became a cultural and tourist attraction, creating a positive spiral of more businesses and new art galleries that settled in the street. In 2010, the municipality initiated a participatory process of refurbishing the street (Genk's spatial-economic vision, 2010). The process was completed halfway through 2014. Adding to the general 'refresher' and beautification of the street, the city built up Vennestraats' image as the 'street of the senses' (Genk's spatial-economic vision, 2010). The idea behind it was to underline the fascinating mix of multicultural catering establishments and shops in the street. More broadly, the city has promoted trade and catering as an asset deployment via a 'max of the mix' marketing strategy based on the multi-'culinarity' of the street [23]. When the refurbishment of the street started, in 2010, 28 shops from a total of 68 were empty. At the moment, there are 76 commercial shops and, an additional one or two available/free spaces—a high occupancy level for such a street [24]. This process was led by the municipality, which established a committee responsible for the image of the commercial street. The committee was formed from representatives of the economic, urban development, and social integration departments and set the leading directions of the refurbishment of the street. This approach further shaped Vennestraat as the main HORECA street of Genk. The first action to be implemented in order to reinvent the street was the Saturday fresh market (in the summer of 2010). The market attracted people in the area, acting as a promoter for the small proprietors. A total renovation of the street, placing urban furniture and greenery followed (works on the street were completed in 2012). The assigned committee developed the aforementioned projects with little to no involvement of the shop owners. The committee organized a call for projects for a vision for the street. The call asked for shop signs, window fronts, and terrace designs, as well as a marketing strategy for Vennestraat (Call for Ideas, 2012). The winning team had to develop, along with the committee, a 'glossary of good practices of interior and graphic designs' to later show proprietors and help them upgrade/update their shops. The objective was clear: a common language for all the shops on the street. In parallel, different actions were organized, by the same committee, to link the street with C-mine and fade out the hierarchy between the two. While C-mine was becoming a strong cultural attractor, Vennestraat would thrive, too. The strategy would benefit both: people that would visit C-mine would pass by Vennestraat, and vice versa. The two function as a 'gastro-cultural' mechanisms in a symbiotic manner. Even though this approach managed to resuscitate the street, with an impact at both the neighborhood and city scale, the participatory process put forward by the municipality created a robust gentrification process. Renting prices rocketed, and smaller proprietors were pushed out—7 small businesses, from the initial 28 when the refurbishment started, had to change location as they could no longer afford renting a space on Vennestraat.

In the new refurbishment dynamic, shop owners have to follow the direction of the committee in most matters of the shop (i.e., the signs on the window front, the graphics of the menu, the way the window front is displayed). Proprietors organized in an association; this, too, was established at the idea of the municipality. Albeit they run each decision that affects proprietors by this association, the final decision goes to the committee. When a shop owner has a new idea or intent to do something with their shop, he or she has to consult with the committee. If the idea is in line with the overall strategy (the one developed by the municipality and the winning team), the committee will approve. If not, it will be subject to change to fit in the predefined discourse. This is likely to exclude one's skills and freedoms in a time when designing for cities is becoming more and more people-orientated. It is this gap that the Capability Approach covers—the framework focuses on both skills and structures and, as such, we use it to (1) develop a framework of analysis to understand the impact of enabling tools on the participation of entrepreneurs in a design process and (2) research how this 'gentrification process' increased the 'freedom' of entrepreneurs to perform the actions that they value versus the perspective of the municipality: how free should these entrepreneurs be (given that there are also other actors with other values and needs involved in the process)?

4. Mapping the Invisible: A Capacity Building Process

In what follows, we adopt Horellis' [3] stance on participation as not an isolated event but a constant communication between different groups that can be assured by using different methods. We apply the Capability Approach to frame a capacity-building process and understand how this framework can support a collective of entrepreneurs to become aware of their capabilities (and the impact of an ongoing urban renewal process on these capabilities).

When addressing complex urban projects, such as the refurbishment of the Vennestraat street, Genks' main economic ecosystem, the variety of stakeholders that is required has a direct impact on the quality of the project, the budget and the power to speed it up or slow it down [25]. In order to engage in a debate with these stakeholders on the given complex societal processes, we use a variety of enabling tools to support a communicative transaction between the parties involved. As stated previously, we frame enabling tools within Horellis' [3] description, as methodologies that support constant dialogue and knowledge exchange between different (groups of) actors.

To better understand local dynamics and methodologies used as enablers of dialogue between different actors, we turned to professionals and practitioners interviews as a first step of the capacity-building process. The preliminary research phase targeted existing methods used to engage entrepreneurs in a debate and support dialogue between stakeholders. Over three months, we conducted a set of thirteen interviews in two steps: (1) an online questionnaire (see Appendix A) set out to introduce researchers to the interviewees, and vice versa, and (2) semi-structured one-to-one interviews (see Appendix B), during which we further detailed on the particularities of the existing enabling tools and their success a/o limitations. We discussed with ten professionals that use participatory methodologies to support various local projects initiated by Genk's municipality a/o bottom-up initiatives on Vennestraat. Additionally, three representatives of the local administration office were interviewed to get a grasp on the officials' stand regarding the use of various enablers of engagement and knowledge transfer. Questions from the interviews were structured around five main topics: (1) the content of the participatory processes, (2) participatory methods used, (3) who coordinates the said processes, (4) who takes part in the processes, and (5) implementation/impact.

Interviews with city representatives disclose a complex administrative structure responsible for assuring a dialogue between civil society and officials. This structure—the Neighborhood Management Department—was established over two decades ago, as a way to address segregation, crime, and inequality in the city by involving people in the decision-making process. Each neighborhood was appointed a manager (wijk manager). The managers' responsibility is to know what is happening in each neighborhood and keep in touch with the people that live there and inform them about future projects and intensions that the administration has for that specific area, and vice-versa: inform the municipality about people's needs, interests, and desires regarding the environment they live in, as well as about their opinions related to the different proposals made. Neighborhood managers use different methodologies to involve people in this process (e.g., flyers, brainstorm techniques, but mostly through direct dialogue where people are informed about the date and the place when a debate will take place).

Practitioners' interviews outline a somewhat different angle on the methods used to engage with locals. They stress political, technical, and critical factors when addressing the dialogue between actors: political factors—the belief that people have the right to participate in processes of urban transformations and expecting that there is some power shifting going on between stakeholders; technical and aesthetical factors—the belief that the spatial solutions that come from these processes could be both aesthetically and technically more challenging and surprising because there is often the belief that all expertise lies in a designer's hands; the critical factor—confronting yourself as a professional with different people you come to engage in a critical debate.

As observed in Table 1, the enabling tools applied so far are focussing on customary and set up strategies that the facilitators are comfortable with: various brainstorm techniques, focus groups and workshops and information meetings. The methods are lined up with the span of the process that range from a few single gatherings to processes that are engaging several months (and sometimes more). These methods and tools can be composed in five general categories:

- Tools for surveying and mapping the existing situation (e.g., social media monitoring, face-to-face surveys).
- Tools for providing information about the progress of the project under the form of one way communication (e.g., via social media, the local press, policy documents and reports).
- Tools for collecting knowledge and gathering feedback through discussion about existing plans for the refurbishment of the street (e.g., neighborhood manager meetings with focus groups, workshops).
- Productive tools—tools that allow for proposing alternative solutions (co-design workshops).
- Decision-making tools allowing proprietors to vote on project proposals.

In order to support entrepreneurs in understanding their capabilities and the impact projects may have on these capabilities, we propose a set of three enabling artefacts that would support *choice, ability, and opportunity*. Table 2 illustrates how the Capability Approach language shaped the capacity-building process using the three different artefacts.

Table 1. Summary of applied enabling tools in the refurbishment process of Vennestraat.

Capability Approach Conversion Factors		Existing Functionings/Enabling Tools	
Choice provision of alternative participatory mechanisms essential to improve democracy among citizens and enable them to participate actively	Content of the Process	How is public administration represented in the process?	Municipal Level is involved and well represented or overrepresented in the process Regional Scale is represented Supra-regional level is underrepresented to missing
		How is public administration involved in particular?	Public Servants as facilitators (experienced in facilitating methods and brainstorming techniques)
		Does administration fund/sponsor the process?	Public funding for neighbourhood management and process
		Who is launching the content/subjects of different processes?	Municipality, Consortium of organizational/institutional partners, activist groups/initiatives Individual Actors, Local Associations/Private Market Parties
	Participatory Methods		
	Methods	Which methods are applied and facilitated?	Focus on traditional methods: focus groups, workshops/brainstorming techniques, meetings/discussion rounds, extended by Social Media Platforms
	Level of participation	Level of participation in general and in particular project	Focal Points: Information - Consultation - Placation - Partnership
	Capacity	Are the participants able to express their interests?	Yes
		Does the process include making proposals?	Yes
	Process design	Who decides on the usage of methods/tools?	Neighbourhood Management (leadership)

Table 1. *Cont.*

Capability Approach Conversion Factors			Existing Functionings/Enabling Tools	
Ability (1) how individuals engage in the deliberation process and (2) their relationship with different actors involved in this process	Coordination	Agreement	How far is the processes politically accepted?	High: Urban Scale: political commitment
			How intensively is the political domain involved in the process?	Intensively: a process that grew in the past 20 years due to the recognition from the political domain of its importance
			Are the involved/responsible political actors committed to the results?	Yes
		Leadership & Integration	Who is leading the process?	City administration Neighbourhood scale
			Is the process linked to other initiatives?	Yes Wijk Management is linked to other neighbourhoods and different departments, Wijk Management (as intermediary position between municipality - neigbourhood - citizens)
		Resources	Has the process necessary resources (money, room, etc.)?	No
			Are there resource restrictions on participant level?	Time, Knowledge, Language Barriers, Cultural Restrictions
		Design	Are the processes designed? Was there a deliberate design process?	Partly
	Participants	Methods	How are the participants chosen? (democratically)	Open for everybody to join: Invitation Letter to all inhabitants Social Media, newspapers, direct approach/invitation by Wijk Managers
		Extent	Is there stability of the number of participants over time?	Enthusiasm in the beginning - fluctuating throughout the process, people dropping out
		Diversity	Age Groups	Adults & working population age group: well represented 25-64 years Representation of young adults in a transition zone Teens and children: underrepresented Elderly people (65+): tendency for overrepresentation
			Is there gender equality in the process?	Male: well represented - overrepresented Female: represented - underrepresented
		Communication	How is communication organised? (within the process)	Initiative: social media, professional networking, personal meetings Leadership: meetings

Table 1. *Cont.*

Capability Approach Conversion Factors			Existing Functionings/Enabling Tools	
Opportunity to engage in the deliberation phase of the capacity building process in order to assess the effectiveness of participation	**Implementation/ Impact**	Influence	Are there plans/designs/actions produced?	Yes
			Were the results implemented in policy, action, program so far?	Urban Scale, used as policy guideline
			To whom are they addressed?	Alderman and Public Administration
			Are there documents with the results of the process? Could the participants influence those documents?	Participants ideas were incorporated in the 'Genk in Sight: Future Scenarios of Genkenaars on their City' report. Will take initiatives to stimulate and integrate the ideas in the policy and bring them to live
		Learning	Are there training sessions foreseen?	Yes
		Implementation	Have the results been implemented?	Partly (smaller actions) the main part is in the initial phase

Table 2. Summary of interactions between the Capability Approach language, the capacity building process, and enabling artefacts.

Capability Approach Conversion Factors	Capacity Building Process	Enabling Tools/Artefacts
Choice provision of alternative participatory mechanisms essential to improve democracy among citizens and enable them to participate actively	The provision of alternative artefacts is crucial to enhance proprietors' freedom in participating in planning processes. What type of tools can proprietors use to engage in the planning process of the shopping street?	**serious games**—structured tools, complemented by **participatory mapping** and **mental images** as unstructured tool that support people in the endeavor of understanding their conversion factors
Ability (1) how individuals engage in the deliberation process and (2) their relationship with different actors involved in this process	(how) Are entrepreneurs engaged in the process of deliberation? What is their relations with outside actors (i.e., architects, engineers, policy makers etc.)? What are entrepreneurs' diverse capacities to appropriate the space of the street a/o urban interventions? (how) Can the capabilities and ties of each proprietor be identified, enhanced and later used to feed a participatory process? Can we talk about collective capacities (i.e., generate collective action, collectively appropriate, change, maintain or improve the built space of Vennestraat)? (How) can we capture appropriate social narratives as data providers in developing a capacity-building process?	**participatory mapping** to draw entrepreneurs' networks and their dynamics **mental images** to capture social narratives and entrepreneurs' perception of their surroundings **serious games** as environments that present a simplified version of the spatial transformation processes taking place in ones' surrounding
Opportunity to engage in the deliberation phase of the capacity building process, under the role of partners, informants, or actors of change in order to assess the effectiveness of participation	Social, economic, and political processes shaping proprietors' freedom to take part in the process of deliberation and appropriate outcomes of the capacity building process. How far does participation go? Does it build on networks among different (groups of) proprietors and expand room for maneuvering to influence or change policy-making?	**participatory mapping** to analyze the potential of proprietors' connections to shape the urban space **mental images** seize proprietors' personal conversion factors' **serious games** as catalysts in building/strengthening communities and creating interactions

4.1. Participatory Mapping

The way people choose to portray their networks can take form in creative methods, such as hand drawn maps of social connections, or 'relational maps' [26]. These maps establish new ways of social networking examination and improve the comprehension upon the way people choose to display their networks [23]. Bagnoli [26] did such an experiment by offering participants pen and paper and later asking them to portray meaningful relationships by placing themselves in the middle of the map and taking into consideration that the distance from the center to each connection showed the strength of that particular connection. As such, she allowed participants to build personal representations in their own ways, providing 'basic scaffolding' [27]. Roseneil [28] conducted similar research and asked participants during interviews to create relational maps. They were to place their social connections into concentric circles. Here, too, the distance from the center showed the degree of importance of that particular connection. The work of Roseneil [28] is mostly in the therapeutic psychology field (psychosocial dynamics), but there are several analogous objectives, with participants asked to build visual patterns to illustrate their relationships, all with the aim of reaching hidden understandings [23]. Following Ingold's [29] correspondence, we proposed a participatory mapping activity asking proprietors to engage with physical materials (e.g., rubbers and pins) and arrange them into correspondence with one another.

4.1.1. Setting

We led a six-month immersive, participatory mapping activity with proprietors from Vennestraat, shop owners that have their businesses on the commercial street. The main interest here was to identify the 'socio-environmental conversion factors'. This participatory mapping exercise provided us with the lens to disclose a detailed landscape of the socio-economic dynamics in Vennestraat. On an operational level, we integrated a visualization tool in a mapping activity to (1) support people in illustrating their networks and the aspects they want to show and (2) to analyze the potential of their connections to shape the urban space and the dynamics on the street [23]. Sensitive to the tensions revealed in the interviews, we organized individual sessions with each proprietor to create the perfect environment where different dynamics may surface. Each shop owner was asked to create a visual portrait of the socio-environment (in this case economic) network of his/her commercial space using an A3 foam board, rubber bands, and pins (Figure 2). The pins represent different individuals a/o other shops, while the rubber bands represent the connections the author of the map has with them. The rubber band assists the author to visualize the relation he/she has with the rest of the 'actors': the longer the distance between pins, the more detached the relation.

Figure 2. Visual portraits.

4.1.2. Benefits

This type of process interacts in a language that stems engagement regardless of a person's background and allowed us to enable participation of different shop owners to visualize, reflect, and take action over their data. In relation to the overall goal of supporting the capacity-building process, the participatory mapping exercise incentivized reflections on the ability of entrepreneurs to identify their networks guided by the question of *(how) can the capabilities and ties of each proprietor be identified, enhanced and later used to feed a participatory process?* These exercises made visible certain freedoms proprietors have: the freedom to choose shops they want to collaborate with, freedom to choose their providers, and freedom to choose the merchandise they want to see. The activity revealed hidden dynamics of networks and ties shop owners were not necessarily aware of, such as an overall link on how shops function at the street level.

4.2. Mental Images

We conducted this artefact-centred description by introducing mental images as an answer to the following operational question: *(How) can we capture appropriate social narratives as data providers in developing a capacity-building process?* Lynch [30] advanced three components that are critical for the analysis of mental images. They include identity, structure, and meaning. Identity involves the extent by which an object or space can be recognized as different from others, as a separate entity. Structure involves the spatial or pattern distinction which an object or space has to other observers or objects. Meaning refers to the practical and emotional sentiment that an individual has for space or an object. Mental images seize people's 'personal conversion factors', as described by Robeyns [15], and become important for they develop into mental maps that capture people's perception of their surroundings [30–33].

Mental maps are requisite in the context of urban identity formation, which is, in part, representative of one's valuation of a places' cultural significance. Fundamentally, mental maps may be formed either through direct or indirect experiences. However, it is essential to acknowledge that the quality of one's visualization about a particular urban space could be distorted if one solely depends on indirect experiences. For instance, media coverage about an event could affect one's perception. Consequently, the likelihood of forming a distorted image increases and therefore undermines the cultural, social, political, or even geographical significance of the urban space in question. Thus, it becomes imperative for individuals to acknowledge that mental maps become more complex over time due to the incorporation of information derived from both direct and indirect experiences. In this case, the image or reputation of urban spaces is best defined through direct experiences, which are quintessential in the development of an impartial evaluation of an urban area.

4.2.1. Setting

For six months, we gathered mental images people have about the street, interviewing more than 120 people (passers-by, residents, and shop owners). We asked participants to portray the most treasured memories they have about Vennestraat, iconic moments a/o places on the street, and reasons they have to return over and over again to this space. The discussion was structured around six main guiding questions: (1) How would you describe the street to someone who doesn't know it? what are the key characteristics? (2) What is the story of the street: anything changed in time? (do you remember stories/events, anecdotes about the street?), (3) Which are the most important shops (on the street) for you, in your eyes, and why? (4) What do you like? What do you dislike about the street?, and (5) What would you like to see happen in the future? As with the participatory mapping, we used an interactive participatory method. We gave each participant a cardboard contour through which he/she could look at the places they were talking about and 'frame' them as a mental image (Figure 3). The given definition of a mental image is that it "[…] is a unique, personal and selective representation of reality" [34]). Based on this definition, it becomes evident that the images are purposed to project an individuals' understanding or interpretation of the significance of a specific urban space. As such, collective mental images may hinder an individual to achieve its dreams when these dreams do not comply with the collective image. The use of these mental images is significant to recognize crucial information for the capacity-building process, data that would have not been otherwise revealed, such as the way Vennestraat street is perceived, based on ones' either objective or subjective experience.

Figure 3. Mental images.

4.2.2. Benefits

As far as the perception of Vennestraat is concerned, the complexity of mental images is yet another key insight. The selective representation of proprietors predicates the implied meaning and significance of this street. The role that direct and indirect experiences played in helping individuals form an image of the Vennestraat demonstrates that valuation is largely volatile. Therefore, personal values and experiences transcend the implied or intended effect of this space. Individuals' experiences might bring about an effect different from that anticipated. Hence, via the use of mental images, the effect of variations in proprietors' understanding and recognition of 'carriers' of identity and psychological boundaries, alongside other typical characteristics of the street, are made visible. Despite the influence of shop owners' unique perception of urban landscape, mental images are a reliable instrument for the capacity-building process. Through mental mapping, insight were acquired in terms of identifying the more collective carriers of urban identity, alongside the general functioning of the street for particular groups. The functioning of the street is limited in that it applies for those proprietors who have similar characteristics. For instance, an area's representation of a group's identity is relative to their collective understanding of the carriers of the foresaid aspect. Therefore, while proprietors construct their own images based on their experiences, mapping these mental images allows for identification of points of convergence and divergence in entrepreneurs' expectations. Mental images are, in themselves, a selective instrument in that they seeks to single out urban elements that play an important role in the capacity-building process.

4.3. Serious Games

Advocates of participatory approaches to planning processes emphasize the potential of the game technique to augment lays' people contribution to research, ease the process of learning, bring about change, guarantee that the methods used are adjusted to people's needs, and eliminate 'designer subjectivity' [35–37]. Serious games are employed within the field of urban planning and policy ever since the 1950s [38,39]. The number of serious games and applications addressing urban matters is continually growing [25]. Serious games and applications came to cover a broad range of themes: serious gaming as a critical thinking mechanism that can assist people in thinking a future development for their cities or find alternatives to known issues [40], games as means to involve citizens in urban matters like sustainable development [41], games as platforms that engage individuals in city planning [42], games as catalysts in building/strengthening communities and creating interactions, and games as means to turn playful experiences into significant memories [43]. Serious games can be one efficient method used in participatory planning as environments that present a simplified version of the spatial transformation processes taking place in ones' surrounding, environments that foster cooperation and understanding [44]. They foster active participation and collaboration with citizens in order to address urban issues and make the transition to a smooth implementation of studied solutions [41,42,44]. Consequently, serious games are safe mediums where people experience and test these transition processes via various scenarios with no real-life consequence. As such, we advanced two serious games (i.e., Floating City and City Makers) to support collaboration and knowledge transfer among the different actors involved in the project of refurbishing Vennestraat street.

4.3.1. Floating City

Floating City is a game-based brainstorming, discussion, and problem-solving activity for small groups. It aims to foster positive thinking, the commonality between participants, and provide a motivating structure for discussions that involves all participants. The game is a digital one (Figure 4). Such games are routinely used to help groups quickly in assigning values and develop hierarchies among these values for a project a/o product or service without getting too caught up with the negativity typically associated with voicing complaints. In Floating City, a specific topic serves as a focus for collective reflection activities. The process of knowledge production by itself, and especially when there is involvement of people in a gathering, becomes a form of mobilization [45]. The mobilization leads to the development of new solutions and actions that are identified, tested, and retested to evaluate their significance in solving particular pitfalls that human beings experience in their day-to-day lives. Knowledge has to be embedded in the cycles involving action-reflection-action over some time [45]. Floating City nurtures the stated process, the nature of action can be made deeper, and thus moves from practicality in problem-solving to more of a social transformation (Table 3).

Table 3. Floating City—game elements.

	Floating City
Name	
Target Group/Age	9+
How many players can play?	1+
Playing Time	5–10 min with 15 min for debriefing
Short Description (Goal, narrative)	A brainstorming game for public spaces in which players can create and publicize their ideas and suggestions for city projects. A balloon is generated for every new idea and slowly elevates the city. Existing balloons can be viewed and rated by others to increase or decrease their effects on the city. All of the generated ideas and their ratings are automatically saved and are available for evaluation after all game sessions.
What do players learn/experience?	Players are proposing ideas, needs, wishes, values, and visions to adapt or improve their city. Each input is represented in a balloon that is carrying the city. Other players can weigh (thumbs up/down) those ideas and add comments. Voting for ideas and clustering them increases insight in the diversity of values among players regarding sustainable futures. A feeling of shared values.
Game rules	The game master or moderator poses a question for the audience. Players then have the possibility to generate their own answers and contributions and to like or comment on the ideas from other players. One round takes as long as the city needs to reach the top of the atmosphere. Then, either a new round starts or another question can be posed. All ideas are represented as balloons that are helping the city fly higher and higher. The more votes a balloon gets, the higher it pulls the city. High balloons then represent shared values. The number of balloons is an indication for the diversity of values.
Equipment/Game material/Interface	The equipment consists of a laptop, monitor/projector, and one keyboard.
Setting (open/closed, suitable for festivals.....)	The game works best in a workshop setting or festival (in this case, it should be used as an idea collector). Playing the game simultaneously with multiple groups may increase the value of the debriefing.
Additional requirements that are necessary to play the game (i.e., tables, quiet environment, necessary space)	There are no additional requirements.
Persons necessary for facilitation (game master)	The game requires a game master to allow for a constructive debriefing.

Figure 4. Floating City.

Setting: We proposed the game to the representatives of the shop association of Vennestraat in its predefined format, the format developed based on the initial findings. We introduced the game to the community as an artefact that can foster action, an action that each proprietor can take regarding the streets' refurbishment. The end goal of the game from the perspective of the capacity-building process, as a popular activity, is to communicate new ideas and information to people, as well as identifying common values, goals and shared dreams.

Playtests: The game was tested three times over four months. The first playtest session had two participants: one representative of the shops association and the street manager. Participants found the digital prototype easy to use and expressed excitement at the possibility of the rest of the community in using it. The first trial was organized in order to acquaint them with the format and methodology. With the help of the street manager, we organized a second playtest, this time only with representatives of the association. All five representatives took part, and the prototype was again well received. Participants found it very simple, with unpretentious graphics that can explain the topic at hand bluntly. A debriefing moment followed the playtest session and was concerned more with the applicability of the game prototype in the daily communication between proprietors, as well as between them and the street manager. The third trial was organized with three of the most active entrepreneurs from the association that take a leading role in the decision-making process. All playtests concluded that such a methodology could ease their communication on given projects and initiatives on the street; however, they would prefer it to be more under the format of an app that one could use at any time from any place, rather than a game that requires people to be in the same room at the same time.

Benefits: The proposed game structures the brainstorming process and provides democratic mechanisms for sharing and evaluating the ideas of others. It promotes reflection as an individual and as a group and reinforces the identification of shared beliefs. Floating City fosters discussion and problem-solving activity for small groups, positive thinking, commonality between participants, and provides a motivating structure for discussions that involves all participants. Using Floating City

4.3.2. City Makers

City Makers is a combined board and card game that focuses on the activation of participants to improve the liveability of their neighborhood or city. This card based game has one simple idea at its core—present projects as a set of steps that people need to collect resources for. By doing so, players will learn about the different resources needed for particular projects happening in the city (Figure 5). The game adapts abstract terms, such as material, permit, and location as resources, which players obtain in the form of cards. Each player receives a project that they need to finish to acquire points. For example, to start a business one might need to have a budget, idea, location, and people to work with. Players can work with or against each other to earn credits by completing projects, investing in each other's projects, or investing in community projects. The player who gains ten points first wins the game. Freire [46] argues that the expression of voices through forms, such as consultation, may lead to the reduction of the awareness that is required to introduce and sustain change in the community. Action alone may bring about the blind action, which may not present the best course of action to get the desired change in the various situations affecting individuals in any given society [46]. The best action, thus, is the one that takes the course of building self-conscious awareness that is coupled with the analysis of the individual's reality. City Makers can act as a medium for participatory action supporting processes that involve reflection, learning, and development of critical consciousness (Table 4).

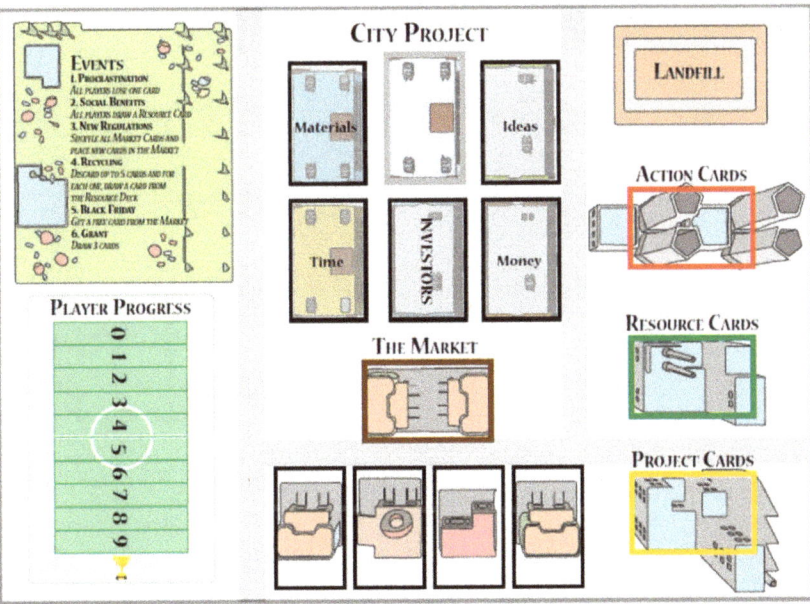

Figure 5. City Makers.

Table 4. City Makers—game elements.

	City Makers
Name	
Target Group/Age	12+
How many players can play?	3–6
Playing Time	30–40 min with 15 min for debriefing
Short Description (Goal, narrative)	Players have to develop projects to improve the livability of a given neighborhood or city. In order to complete these projects, they can either chose to compete or to collaborate. The players can earn credits by completing projects, investing in each other's projects, or investing in community projects. The player who reaches ten points first wins the game.
What do players learn/experience?	The players are asked to present (given) projects as a set of steps that they need to collect resources (material, permit and location) for. By doing so, players will learn about the procedures that are required to implement these projects. This may, in turn, trigger them to actually consider implementing a project.
Game rules	Each player receives a project (in the form of a card) that he/she needs to finish to acquire points. For example, to start a business one might need to have a budget, idea, location, and people to work with. The player, who reaches ten points first wins. The game consists of four sets of cards: project, resource, market, and action cards. They are placed in the predefined deck positions. One project and four market cards are drawn from the decks and placed so everyone can see them. Everyone receives three color tokens—one for representing them on the field and two for investments—, three player cards, and a project. Players are then allowed to draw an additional resource card each turn, trade with everyone, and do only one of the following: finish one step of their project, invest in the common project or another player's project, or play an action card.
Equipment/Game material/Interface	The equipment consists of a (printable) game-board and a (printable) set of cards.
Setting (open/closed, suitable for festivals...)	The game works best in a workshop setting. Playing the game simultaneously with multiple groups may increase the value of the debriefing.
Additional requirements that are necessary to play the game (i.e., tables, quiet environment, necessary space)	There are no additional requirements.
Persons necessary for facilitation (game master)	The game requires a game master and an observant. The observant documents the gameplay to allow for a constructive debriefing.

Setting: After introducing Floating City to the group, we proposed City Makers as a follow-up game that will incorporate ideas gathered from the brainstorming activity under the form of projects and further detail on these projects. The game proposed clear projects, from parking area for Vennestraat, less parking spaces on the street, larger sidewalks, and more sitting spots for visitors on the street, to more abstract ones (e.g., a greener Vennestraat street, how to make the shopping street more attractive, how to maintain its charisma).

Playtests: City Makers was played twice over one month. As with Floating City, we first introduced the game to the street manager and one of the association representatives. It was immediately labeled as a complicated activity and a type of game that only association and city administration representatives could play as it would be too complex for the other shop owners. The first test revolved around how to translate ideas from Floating City to City Makers, the games' mechanics, and narrative. Concerns on the lengthy process the game brings to the table were underlined; participants were reserved regarding the timeframe of the workshop. The second playtest was organized with four association members. Each test took about 30 to 40 min, with 15 min for debriefing. Emergent game situations were observed, and proprietors started roleplaying and established alliances. Shop owners formed strategies depending on the available projects, resources, and emergent social dynamics. Participants traded with each other constantly and invested in other projects more than in their own, with the goal to obtain additional points when other players succeed. The game proved to be equally fun and motivating, and proprietors adapted metaphors used for the resource cards (e.g., people, time, money) to envision fun scenarios where they would be the main actors responsible for one of the resources. They would then develop a storyline where a particular type of exchange would take place and detail on the impact of their imagined actions. At the end of the game, the shop owners were asked to customize a project of their own with available resources. By motivating them to reflect within the constraints of the game setting, they link the process and understand the necessary steps in their real-world projects.

Benefits: The game fosters reflection on the steps needed to take in real life projects by motivating players to think within the constraints of the game. Herein lies the learning aspect of the game—through the abstraction of resources and social interactions, players attain certain ideas about how this can work in real-life scenarios. For the capacity-building process, this particular game brought about interpersonal communication, allowing a/o forcing participants to verbalize and, therefore, more profoundly concern themselves with the projects proposed for the street, their beliefs, and ideas, as well as those from others. Proprietors built capacities to better understand the choices available and necessary steps needed for a specific intervention, as well as the opportunities and abilities they have to participate in implementing these projects.

We would like to acknowledge the small size of the sample during the game session—a total of five play tests ranging from two to five participants per session. However, this does not undermine reported experiential evidence, as it is not structured scientific data. As such, we invite other authors to experiment with this tool and benefit from its versatility.

5. Conclusions

The capacity-building process is a non-linear procedure acting as a medium that interlinks spaces (geographies) and reveals choices, abilities and opportunities relating to Sen's and Nussbaum's notions of freedom. Building capacities for proprietors that will, later on, create the space of the Vennestraat street greets principles outlined by the Capability Approach framework. Ranciere and Corcoran [47] too, strengthen the nature of the capacity-building process as they claim that conceiving a design is not an imposed form. Instead, it is a set of artefacts and practices that result in different relations based on speculative thinking and time. We conclude with the evaluation of the enabling tools used in the Vennestraat capacity-building process (Table 5).

Table 5. Evaluation of the enabling artefacts within the Vennestraat capacity-building process against the Capability Approach language.

Capability Approach Conversion Factors	Capacity Building Process	Enabling Tools/Artefacts
Choice provision of alternative participatory mechanisms essential to improve democracy among citizens and enable them to participate actively	• Analyzing a/o addressing social needs • Experimentation with disruptive tools • Deliberate use of ideas a/o mechanism that seek to challenge the existing practices established in the street • Measures or resources to allow direct involvement of proprietors in participatory processes or independently realize their freedoms to act • Involvement of diverse (groups of) actors in the process	**serious games**—structured tools, complemented by **participatory mapping** and **mental images** as unstructured tool that support people in the endeavor of understanding their conversion factors
Ability (1) how individuals engage in the deliberation process and (2) their relationship with different actors involved in this process	• A variety of actors involved throughout the process • Integrating into the design of the street different opinions voiced by proprietors • Involvement of various and multiple stakeholders in knowledge production processes • Stakeholders reflecting on capacity building processes • Capacity building in relation to building social capital by creating relations of trust between proprietors or formal communication channels for marginalized shop owners that could facilitate future collaboration • Recognizing systemic barriers (e.g., regulations, physical barriers, cultural values) that need to be overcome for various interventions on the street to become viable or successful and • Formulate explicit strategies to overcome such path dependencies	**participatory mapping** to draw entrepreneurs' networks and their dynamics **mental images** to capture social narratives and entrepreneurs' perception of their surroundings **serious games** as environments that present a simplified version of the spatial transformation processes taking place in ones' surrounding
Opportunity to engage in the deliberation phase of the capacity building process, under the role of partners, informants, or actors of change in order to assess the effectiveness of participation	• Leadership acting as a driving collaborative force in the process • Agendas aiming to tackle sustainability challenges • Shape an explicit future shared vision of the street • Project activities contributing to capacity development across human action levels • Strategies seeking to reveal proprietors freedoms and quality of life • Providing entrepreneurs with new skills, training, abilities and improved access to participatory processes tackling the future development of the street • Formulate ideas that are (1) explicit, (2) aim for radical change, and (3) supported by a wide range of proprietors	**participatory mapping** to analyze the potential of proprietors connections to shape the urban space **mental images** seize proprietors' 'personal conversion factors' **serious games** as catalysts in building/strengthening communities and creating interactions

Oosterlaken and Hoven [7] consider participation a design issue, as opposed to being theoretical. In other words, we, as researchers a/o practitioners, ought to consider the capability concept in a speculative term, instead of a normative expression. Latour [48] links norms with the capability space because they stem from the movement of things from "objects" to "matters of concern". As "matters of concern", associates with building arguments for intervention that respond to the speculative, productive, and innovative potentialities. The participatory mapping exercise revealed proprietors networks and supported shop owners in understanding the potential these networks have in shaping the urban space that influences their daily routine. While picturing their mental images, participants capture their perception of the surrounding environment seizing their conversion factors). Complementary to the first two, serious games, even when used in a limited setting, supported interaction among participants and presented the refurbishment process of Vennenstraat in a simplified manner. As such, the different enabling tools (i.e., participatory mapping, mental images, and serious games), engaged in the process of building capacities among proprietors, allowed participants and researchers to grasp the overall institutional setting that influences the transformation factors and one's capabilities in the given context.

This article is a plea for an active inclusion, acknowledgement, and development of capabilities in complex participatory projects among different stakeholders and provides researchers and practitioners alike with a set of artefacts to help engage in said processes. Equally, the article is a shy invitation to start a dialogue on capabilities, considering that the classification presented above is based on a specific context and topic, leaving numerous dimensions of the tools un(der)addressed. We would like to underline the exploratory nature of the capacity building process and encourage other authors to build on such preliminary exploration to gather more evidence and formal results/analyses to deepen the understanding of how proposed ideas work. We invite the readers to experiment with enabling tools to support collectives in understanding the systems that block a/o support their freedoms and understand how they are part of this system and may also block a/o support the freedoms of others.

Author Contributions: The individual contribution of each author of this paper is as follows: Conceptualization, T.I.C. and O.D.; Methodology, T.I.C.; Software, T.I.C.; Validation, T.I.C. and O.D.; Formal Analysis, T.I.C.; Investigation, T.I.C.; Resources, T.I.C. and O.D.; Data Curation, T.I.C.; Writing—Original Draft Preparation, T.I.C.; Writing—Review & Editing, T.I.C. and O.D.; Visualization, T.I.C.; Supervision, O.D.; Project Administration, O.D.; Funding Acquisition, O.D. All authors have read and agreed to the published version of the manuscript.

Funding: This research has received funding from the European Union's Horizon 2020 research and innovation programme under grant agreement No 857160 [JPI Urban Europe grant number 857160].

Conflicts of Interest: The authors declare no conflict of interest.

Appendix A. Online Questionnaire

Participatory Processes in Planning

1. Through your professional occupation, in how many participatory projects were you involved in the last five years?
2. In which field do the projects you are involved with, include participatory processes?
3. Which social groups are over-, underrepresented, or completely missing from the participatory process?
4. Which gender groups are over-, underrepresented or completely missing from the participatory process?
5. Which age groups are over-, underrepresented, or completely missing from the participatory process?
6. How are participation processes initiated?
7. How much influence do the participants have on setting the rules of the participatory processes
8. What's the average duration of participation projects?
9. What's the type of events employed in participation projects?
10. What kind of tools are employed the most in these participatory processes?

Appendix B. Semi-Structured Interview

Introductory questions:

0. Can you briefly explain the aim and goals of your organization and your position?
0. What kind of experience do you have with participatory processes?
0. How do you personally participate? What is your role in the participatory process?

Participatory Processes and Planning:

1. In which areas you consider important to use participation?

- What are the goals of actor involvement in the framework of this particular project?
- Can you describe the participation processes in the projects?
- Which tool(s) did you find as being most productive in terms of the process and quality of outcome? Why/How?
- What is your motivation for using participatory processes?
- What are your expectations?

2. How are people selected or invited to participate? How can and how does one join in?

- What are the incentives for e.g., private persons, who take part in these processes over a period of time?
- How do you reach out to potential participants?

3. Can you describe the participatory process that you are using?

- When does the process stop? How do you know, when it's finished?
- When do people drop out, and who is dropping out of the process?
- Why in your opinion people stop participating?

4. What kind of rules are governing the participation process?

- Who decides on rules? Can they be changed?
- Are participants generally accepting and playing by rules?
- Do participants bend, ignore or break the rules?

5. How do you deal with conflict and power relations that are inherent in the participation process? How do you deal with the power conflict between the formal inviting organization and the participants?

- Which kind of conflicts surface during a participation process?
- How do you deal with power differentials within the group?
- How do you deal with knowledge and time availability differentials within the group?
- How do you deal with people who obstruct the process? (e.g., By bullying other participants, hampering or screaming?)

6. What are results or outcomes of the projects?

- How satisfied are you with the outcomes? Did they match your expectations?
- How much influence do existing participatory processes have on actual planning policies or plans?
- How do you manage expectations on the part of participants?

References

1. Schouten, B.; Ferri, G.; de Lange, M.; Millenaar, K. Games as Strong Concepts for City-Making. In *Playable Cities, Gaming Media and Social Effects*; Springer: Singapore, 2017. [CrossRef]

2. Suchman, L.; Blomberg, J.; Orr, J.E.; Trigg, R. Reconstructing Technologies as Social Practice. *Am. Behav. Sci.* **1999**, *43*, 392–408. [CrossRef]
3. Horelli, L. A Methodology of Participatory Planning. In *Handbook of Environmental Psychology*; Wiley: New York, NY, USA, 2002; pp. 607–628.
4. Frediani, A.A. Participatory Methods and the Capability Approach. Human Development and Capability Association Briefings. Available online: http://www.capabilityapproach.com/pubs/Briefing_on_PM_and_CA2.pdf (accessed on 14 January 2006).
5. Ranciere, J. *Staging the People: The Proletarian and His Double*; Verso: Brooklyn, NY, USA, 2011; ISBN 978-1-84467-697-2.
6. Frediani, A.A. The Capability Approach as a Framework to the Practice of Development. *Dev. Pract.* **2010**, *20*, 173–187. [CrossRef]
7. Oosterlaken, I. Design for Development: A Capability Approach. *Des. Issues* **2009**, *25*, 91–102. [CrossRef]
8. Sen, A. Development as Capability Expansion. In *Readings in Human Development*; Oxford University Press: New York, NY, USA, 2003.
9. Nussbaum, M. Creating Capabilities: The Human Development Approach and Its Implementation. *Hypatia* **2009**, *3*, 211–215. [CrossRef]
10. Sen, A. *The Country of First Boys: And Other Essays*; Oxford University Press: New York, NY, USA, 2015; ISBN 9780198738183.
11. Sen, A. Human Rights and Capabilities. *J. Hum. Dev.* **2005**, *6*, 151–166. [CrossRef]
12. Alkire, S. *Valuing Freedoms: Sen's Capability Approach and Poverty Reduction*; Oxford University Press: Oxford, UK, 2002.
13. Gore, C. Irreducible social goods and the informational basis of Amartya Sen's capability approach. *J. Int. Dev.* **1997**, *9*, 235–250. [CrossRef]
14. Sen, A. *Development as Freedom*; Oxford University Press: Oxford, UK, 1999.
15. Robeyns, I. The Capability Approach: A theoretical survey. *J. Hum. Dev.* **2005**, *6*, 93–117. [CrossRef]
16. Alkire, S. Choosing dimensions: The Capability Approach and Multidimensional Poverty. Available online: https://www.files.ethz.ch/isn/127875/WP88_Alkire.pdf (accessed on 29 January 2020).
17. Madanipour, A. *Whose Public Space? International Case Studies in Urban Design and Development*; Routledge: London, UK; New York, NY, USA, 2010.
18. Goleman, D. *Focus: The Hidden Driver of Excellence*; HarperCollins Publisher: New York, NY, USA, 2013.
19. Innes, E.J.; Booher, D.E. *Planning with Complexity—An Introduction to Collaborative Rationality for Public Policy*; Routledge: London, UK; New York, NY, USA, 2010.
20. Dong, A. The Policy of Design: A Capabilities Approach. *Des. Issues* **2008**, *24*, 76–87. [CrossRef]
21. Ehn, P. Participation in Design Things. In *Proceedings of the Tenth Anniversary Conference on Participatory Design 2008*; ACM: New York, NY, USA, 2008; pp. 92–101.
22. Muller, E. Formatted Spaces of Participation: Interactive television and the changing relationship between production and consumption. In *Digital Material—Tracing New Media in Everyday Life and Technology*; Amsterdam University Press: Amsterdam, The Netherland, 2009; pp. 49–63.
23. Constantinescu, T.; Devisch, O. *Portraits of Work: Mapping Emerging Coworking Dynamics*; Information, Communication & Society: Manchester, UK, 2018.
24. Belgian Statistical Office—Stadbel. Available online: https://statbel.fgov.be/en (accessed on 29 January 2020).
25. Constantinescu, T.; Devisch, O.; Kostov, G. City Makers: Insights on the Development of a Serious Game to Support Collective Reflection and Knowledge Transfer in Participatory Processes. *Int. J. E-Plan. Res.* **2017**, *6*. [CrossRef]
26. Bagnoli, A. Beyond the standard interview: The use of graphic elicitation and arts-based methods. *Qual. Res.* **2009**, *9*, 547–570. [CrossRef]
27. Prosser, J.; Loxley, A. Introducing Visual Methods. Available online: http://eprints.ncrm.ac.uk/420/1/MethodsReviewPaperNCRM-010.pdf (accessed on 29 January 2020).
28. Roseneil, S. The ambivalences of angel's "arrangement". *Sociol. Rev.* **2006**, *54*, 847–869. [CrossRef]
29. Ingold, T. *Making: Anthropology. Archaeology: Art and Architecture*; Routledge: Abingdon, UK, 2013.
30. Lynch, K. *The Image of the City*; MIT press: Cambridge, MA, USA, 1960.
31. Wurman, R.S. Information Anxiety: Towards Understanding. Available online: https://scenariojournal.com/article/richard-wurman/ (accessed on 29 January 2020).

32. Tufte, E. The Visual Display of Quantitative Information. Cheshire Connecticut. Available online: http://www.colorado.edu/UCB/AcademicAffairs/ArtsSciences/geography/foote/maps/assign/reading/TufteCoversheet.pdf (accessed on 29 January 2020).
33. Amoroso, N. *The Exposed City: Mapping the Urban Invisible*; Routledge: London, UK, 2010; pp. 41–64.
34. Sulsters, W. Mental Mapping: Viewing the Urban. Landscapes of the Mind. Available online: http://resolver.tudelft.nl/uuid:fc71de16-b485-4888-b6fe-a9d2771d9e4a (accessed on 29 January 2020).
35. Rogers, Y.; Marsden, G. Does he take sugar? Moving beyond the rhetoric of compassion. *Interactions* 2013, *20*, 48–57. [CrossRef]
36. Vines, J.; Clarke, R.; Wright, P.; McCarthy, J.; Olivier, P. Configuring Participation: On How We Involve People in Design. In Proceedings of the SIGCHI Conference on Human Factors in Computing Systems, Paris, France, 27 April–2 May 2013; ACM: New York, NY, USA, 2013; pp. 429–438.
37. Muller, M.J. *The Human-Computer Interaction Handbook. Chapter Participatory Design: The Third Space in HCI*; L. Erlbaum Associates Inc.: Hillsdale, NJ, USA, 2003; pp. 1051–1068.
38. Abt, C. *Serious Games*; Viking Press: New York, NY, USA, 1969.
39. Duke, R. *Metropolis: The Urban Systems Game*; Gamed Simulations, Inc.: New York, NY, USA, 1975.
40. Flanagan, M. *Critical Play: Radical Game Design*; MIT Press: Cambridge, MA, USA, 2009.
41. De Lange, M. ECLECTIS Report: A Contribution from Cultural and Creative Actors to Citizens' Empowerment. Available online: https://waag.org/sites/waag/files/media/publicaties/publication_eclectis_bd.pdf (accessed on 29 January 2020).
42. Tan, E.; Portugali, J. The Responsive City Design Game. In *Complexity Theories of Cities Have Come of Age*; Springer: Berlin, Germany, 2012; pp. 369–390.
43. Rieser, M. Locative voices and cities in crisis. Studies in documentary film. In *Playable Cities: The City as a Digital Playground*; Springer: Singapore, 2012.
44. Gordon, E.; Baldwin-Philippi, J. Playful Civic Learning: Enabling Reflection and Lateral Trust in Game-based Public Participation. *Int. J. Commun.* 2014, *8*, 759–786.
45. Ashiem, B.; Cohen, L.; Vang, J. Face-to-face, buzz and knowledge bases: Socio-spatial implications for learning, innovation and innovation policy. *Environ. Plann. C* 2007, *25*, 655–670. [CrossRef]
46. Freire, P. *Cultural Action for Freedom*; Harvard Education Press: Cambridge, MA, USA, 1970.
47. Ranciere, J.; Corcoran, S. *Dissensus on Politics and Aesthetics*; Continuum London: London, UK, 2010.
48. Latour, B. Why Has Critique Run out of Steam? From Matters of Fact to Matters of Concern. *Crit. Enq.* 2004, *30*, 225–248. [CrossRef]

© 2020 by the authors. Licensee MDPI, Basel, Switzerland. This article is an open access article distributed under the terms and conditions of the Creative Commons Attribution (CC BY) license (http://creativecommons.org/licenses/by/4.0/).

MDPI
St. Alban-Anlage 66
4052 Basel
Switzerland
Tel. +41 61 683 77 34
Fax +41 61 302 89 18
www.mdpi.com

Informatics Editorial Office
E-mail: informatics@mdpi.com
www.mdpi.com/journal/informatics

www.ingramcontent.com/pod-product-compliance
Lightning Source LLC
LaVergne TN
LVHW070720100526
838202LV00013B/1135